The Sassafras Guide to Earth Science

Written by Paige Hudson

The Sassafras Guide to Earth Science

First Edition 2016
Copyright @ Elemental Science, Inc.
Email: info@elementalscience.com

ISBN # 978-1-935614-45-6

Printed In USA For World Wide Distribution

For more copies write to :
Elemental Science
610 N. Main St. #207
Blacksburg, VA 24060
info@elementalscience.com

Copyright Policy

All contents copyright © 2016 by Elemental Science. All rights reserved.

No part of this document or the related files may be reproduced or transmitted in any form, by any means (electronic, photocopying, recording, or otherwise) without the prior written permission of the author. The author does give permission to the original purchaser to photocopy all supplemental material for use within their immediate family only.

Limit of Liability and Disclaimer of Warranty: The publisher has used its best efforts in preparing this book, and the information provided herein is provided "as is." Elemental Science makes no representation or warranties with respect to the accuracy or completeness of the contents of this book and specifically disclaims any implied warranties of merchantability or fitness for any particular purpose and shall in no event be liable for any loss of profit or any other commercial damage, including but not limited to special, incidental, consequential, or other damages.

Trademarks: This book identifies product names and services known to be trademarks, registered trademarks, or service marks of their respective holders. They are used throughout this book in an editorial fashion only. In addition, terms suspected of being trademarks, registered trademarks, or service marks have been appropriately capitalized, although Elemental Science cannot attest to the accuracy of this information. Use of a term in this book should not be regarded as affecting the validity of any trademark, registered trademark, or service mark. Elemental Science is not associated with any product or vendor mentioned in this book.

The Sassafras Guide to Earth Science
Table of Contents

Introduction...5

Book List...7

Microscope Information...10

Demonstration Supplies Listed By Chapter...11

Project and Activity Supplies Listed By Chapter....................................13

The Sassafras Guide to the Characters...15

Chapter 1: Embarking on Earth Science..19

Chapter 2: O-o-o-o-klahoma..23

Chapter 3: Lucille's First Rodeo..29

Chapter 4: The Congolese Jungle Treasure Hunt...................................33

Chapter 5: The Search for the Giant Bonobo Diamond............................37

Chapter 6: Parachuting into Patagonia..41

Chapter 7: Out of the Office..45

Chapter 8: The Gobi Desert...49

Chapter 9: Avargatom Challenges..53

Chapter 10: Wolves in Pakistan..57

Chapter 11: The Lost One is Found..61

Chapter 12: Back in Alaska...65

Chapter 13: The Forget-o-nator..69

Chapter 14: A Watery Landing . . . Again..73

Chapter 15: The Threat of Thaddaeus...77

Chapter 16: Quick! To Switzerland!...81

Chapter 17: Tracking Down Bogdanovich......................................85

Chapter 18: The End of Earth Science..89

Appendix..93

Lab Report Sheet	*95*
Microscope Worksheet Template (Older)	*97*
Microscope Worksheet Template (Younger)	*98*
Weather Pictures	*99*
The Water Cycle (Blank)	*100*
The Water Cycle (Completed)	*101*
The Nitrogen Cycle (Blank)	*102*
The Nitrogen Cycle (Completed)	*103*
The Phosphorus Cycle (Blank)	*104*
The Phosphorus Cycle (Completed)	*105*
Map Template	*106*

Glossary...107

Quizzes...111

Earth Science Quiz Answers	*112*
Earth Science Quiz #1	*113*
Earth Science Quiz #2	*115*
Earth Science Quiz #3	*117*
Earth Science Quiz #4	*119*
Earth Science Quiz #5	*121*
Earth Science Quiz #6	*123*
Earth Science Quiz #7	*125*
Earth Science Quiz #8	*127*

The Sassafras Guide to Earth Science Introduction

Our Living Books' method of science instruction was first proposed in *Success in Science: A Manual for Excellence in Science Education*. This approach is centered on living books that are augmented by notebooking and scientific demonstrations. The students read (or are read to) from a science-oriented living book, such as *The Sassafras Science Adventures Volume 4: Earth Science*. Then, they write about what they have learned and complete a related scientific demonstration or hands-on project. If the time and interest allow, the teacher can add in non-fiction books that coordinate with the topic, or the students can do an additional activity and memorize related information.

The books of the *Sassafras Science Adventures* series are designed to give you the tools you need to employ the Living Books' method of science instruction with your elementary students. For this reason, we have written an activity guide and logbook to correspond with each novel. This particular activity guide contains eighteen chapters of activities, reading assignments, scientific demonstrations, and so much more for studying earth science.

Each of the chapters in this guide corresponds directly with the chapters in *The Sassafras Science Adventures Volume 4: Earth Science*. They are meant to give you the information you need to turn the adventure novel into a full science course for your elementary students. The chapters will provide you with a buffet of options that you can use to teach your students about weather, oceans, and more. So pick and choose what you know you and your students will enjoy!

What Each Chapter Contains

Each chapter begins with a summary of the corresponding chapter in *The Sassafras Science Adventures Volume 4: Earth Science*. Then, there will be an overview of the supplies you will need for the demonstration, projects, and activities for the chapter. After that, you will find the optional schedules – one for two days a week and one for five days a week. These schedules are included to give you an idea of how your week could be organized, so please feel free to alter them to suit your needs.

After the week-at-a-glance information, you will find the information for the reading, notebooking, and activities for the particular chapter. This information is divided into the following sections:

Science-Oriented Books
- **Chapter Summary** – This section contains a paragraph summary of the corresponding chapter in *The Sassafras Science Adventures Volume 4: Earth Science*.

- **Encyclopedia Readings** – This section contains possible reading assignments from:
    - *Basher Science Planet Earth* (best for 1st through 2nd grades as a read aloud)
    - *Usborne Children's Encyclopedia* (best for 2nd through 4th grades)
    - *Discover Science Weather* (best for 2nd through 4th grades)
    - *Usborne Encyclopedia of Planet Earth* (best for grades 4th through 6th grades)

    You can choose to read the assignments to the students or have the students read them on their own.

- **Additional Living Books** – This section contains a list of books that coordinate with what is being studied in the chapter. You can check these books out of your local library.

Notebooking
- **SCIDAT Logbook Information** – This section has the information that the students could include in their SCIDAT logbook. It contains possible earth science information the student could

include on their earth science record sheets. The students may or may not have all the same information on their notebooking sheets, which is fine. You want their SCIDAT logbook to be a record of what they have learned. The information included is meant for you to use as a guide as you check their work. For more information about notebooking, please read the following articles:
- What is notebooking? – http://sassafrasscience.com/what-is-notebooking/
- How to use notebooking with different ages – http://sassafrasscience.com/notebooking-with-different-ages/

✧ VOCABULARY – This section includes vocabulary words that coordinate with each chapter. If your students are older, I recommend that you have them create a glossary of terms using a blank sheet of lined paper or the glossary sheets provided in *The Official Sassafras Student SCIDAT Logbook: Earth Science Edition*. You can also have them memorize these words and their definitions.

Scientific Demonstrations or Observations
☑ SCIENTIFIC DEMONSTRATION – This section includes a list of materials, the instructions, and an explanation for a scientific demonstration that coordinates with the chapter. A blank lab report sheet is provided for you in the Appendix on pp. 96-96 if you wish your students to write up the demonstration. If your students are in fourth grade or higher, I recommend that they complete at least one of these lab reports for this course.

Multi-week Projects or Activities
✂ ADDITIONAL ACTIVITIES – This section contains additional activities that go along with the chapter. There are multi-week projects, which will be done over several chapters, and activities that coordinate with that specific chapter. Pick and choose the activities that interest you and your students.

Memorization
☞ COPYWORK AND DICTATION – This section contains a short copywork passage and a longer dictation passage for you to use. Some students may use the shorter passages for dictation or the longer passages for copywork. Feel free to tailor the selections to your students' abilities. You can also use the selections as memory work assignments for the students.

Additional Materials

In the back of this guide, there are a few additional materials for your convenience. The first is a glossary of terms, which you can use with your students as they define the words for each chapter. After that, you will find a set of eight simple quizzes that you can use with your students to verify if you students are retaining the material.

A Word About the SCIDAT Logbook

The SCIDAT logbook is meant to be a record of your students' journey through their study of earth science. It is explained in more detail in Chapter 1 of this guide. You can choose to make your own or purchase a pre-made logbook from Elemental Science. *The Official Sassafras SCIDAT Logbook: Earth Science Edition* has all the pages the students will need to create their own logbooks. Each one has been attractively illustrated for you so that you don't have to track down pictures for the students to use. This way, the students are able to focus on the information they are learning.

Final Thoughts

As the author and publisher of this curriculum, I encourage you to contact me at info@elementalscience.com with any questions or problems that you might have concerning *The Sassafras Guide to Earth Science*. I will be more than happy to answer them as soon as I am able. I hope that you and your students enjoy your journey through the world of Earth Science with the Sassafras twins.

Book List

Main Text

The following book is required reading for the activities suggested in this guide.

- *The Sassafras Science Adventures Volume 4: Earth Science*

Encyclopedia Readings

The following encyclopedias have suggested pages scheduled in this guide. I recommend that you choose the one that best suits the age and ability of your students.

- *Basher Science Planet Earth* (best for 1st through 2nd grades as a read aloud)
- *Usborne Children's Encyclopedia* (best for 2nd through 4th grades)
- *Discover Science Weather* (best for 2nd through 4th grades)
  (NOTE – The Discover Science Weather book has been known to go in and out of print. However, it is still one of the best options for elementary students, which is why we included it as an option.)
- *Usborne Encyclopedia of Planet Earth* (best for grades 4th through 6th grades)

Recommended Resources

The following book will be very beneficial to have when completing this course. It contains all the pages and pictures your students will need to record their journey through Earth Science.

- *The Official Sassafras Student SCIDAT Logbook: Earth Science Edition*

View all the links mentioned in this guide in one place and get a digital copy of the templates, glossary, and quizzes by visiting the following page:

- http://sassafrasscience.com/volume-4-links/

Additional Living Books Listed By Chapter

Chapter 1
- *On the Same Day in March: A Tour of the World's Weather* by Marilyn Singer and Frane Lessac
- *Weather and Climate: Geography Facts and Experiments* (Young Discoverers Series) by Barbara Taylor

Chapter 2
- *Wind* by Marion Dane Bauer and John Wallace
- *Feel the Wind* (Let's-Read-and-Find... Science 2) by Arthur Dorros
- *The Wind Blew* by Pat Hutchins
- *Like a Windy Day* by Frank Asch

Chapter 3
- *Tornadoes!* by Gail Gibbons
- *Tornadoes* by Seymour Simon
- *Tornado Alert* (Let's-Read-and-Find-Out Science 2) by Franklyn M. Branley and Giulio Maestro
- *A Grassland Habitat* (Introducing Habitats) by Kelley Macaulay and Bobbie Kalman
- *Grasslands* (About Habitats) by Cathryn P. Sill

CHAPTER 4
- *Down Comes the Rain* (Let's-Read-And-Find... Science 2) by Franklyn Mansfield Branley
- *The Rain Came Down* by David Shannon
- *Rain* (Weather Series) by Marion Dane Bauer and John Wallace
- *A Rainforest Habitat* (Introducing Habitats) by Molly Aloian

CHAPTER 5
- *Flood* (Capstone Young Readers) by Alvaro F. Villa
- *National Geographic Readers: Storms!* by Miriam Goin
- *Wild Weather, Level 1 Extreme Reader* (Extreme Readers) by Katharine Kenah

CHAPTER 6
- *Snow* by Uri Shulevitz
- *Snow Is Falling* (Let's-Read-and-Find... Science, Stage 1) by Franklyn M. Branley and Holly Keller
- *Snow* (Ready-to-Reads) by Marion Dane Bauer and John Wallace

CHAPTER 7
- *Tree For All Seasons* (Avenues) by Robin Bernard
- *On the Same Day in March: A Tour of the World's Weather* by Marilyn Singer and Frane Lessac
- *Sunshine Makes the Seasons* (Let's-Read-and-Find... Science 2) by Franklyn M. Branley and Michael Rex
- *Watching the Seasons* (Welcome Books) by Edana Eckart
- *The Reasons for Seasons* by Gail Gibbons

CHAPTER 8
- *Day and Night* (First Step Nonfiction: Discovering Nature's Cycles) by Robin Nelson
- *Day and Night* (Patterns in Nature) by Margaret Hall and Jo Miller
- *What Makes Day and Night* (Let's-Read-and-Find... Science 2) by Franklyn M. Branley and Arthur Dorros
- *Extreme Weather: Surviving Tornadoes, Sandstorms, Hailstorms, Blizzards, Hurricanes, and More!* by Thomas M. Kostigen

CHAPTER 9
- *Droughts* (Weather Update) by Nathan Olson
- *Droughts* (Blastoff Readers Level 4) by Anne Wendorff
- *A Desert Habitat* (Introducing Habitats) by Kelley Macaulay and Bobbie Kalman
- *About Habitats: Deserts* by Cathryn P. Sill
- *Life in the Desert* (Pebble Plus: Habitats Around the World) by Alison Auch

CHAPTER 10
- *Grandpa, What is Air?* (Popular Science for Children) by Daniel Levy, Efraim Perlmutter, and Yona
- *Atmosphere: Air Pollution and Its Effects* (Our Fragile Planet) by Dana Desonie
- *Little Cloud* by Eric Carle

CHAPTER 11
- *Clouds* (Let's-Read-and-Find... Science 1) by Anne Rockwell and Frane Lessac
- *The Cloud Book* by Tomie dePaola
- *Clouds* (Ready-to-Reads) by Marion Dane Bauer and John Wallace

CHAPTER 12
- *The Water Cycle* by Helen Frost
- *The Water Cycle* (Earth and Space Science) by Craig Hammersmith

- *The Water Cycle (Water All Around)* by Rebecca Olien and Ted Williams
- *Fog and Mist (Watching the Weather)* by Elizabeth Miles
- *Fog (Weather)* by Helen Frost

CHAPTER 13
- *The Nitrogen Cycle (Cycles in Nature)* by Suzanne Slade

CHAPTER 14
- *National Geographic Readers: Coral Reefs* by Kristin Rattini
- *Exploring Coral Reefs* by Anita Ganeri
- *Coral Reefs* by Seymour Simon
- *Over in the Ocean: In a Coral Reef* by Marianne Berkes

CHAPTER 15
- *Eye Wonder: Oceans (DK Eyewonder)*
- *Basher Science: Oceans: Making Waves!*
- *Hurricane Watch (Let's-Read-and-Find-Out Science 2)* by Melissa Stewart and Taia Morley
- *Hurricanes (Little Scientist)* by Martha E. H. Rustad
- *The Magic School Bus Inside A Hurricane* by Joanna Cole

CHAPTER 16
- *Future Engineering: The Clean Water Challenge* by Robyn Friend
- *John Muir Wrestles a Waterfall* by Julie Danneberg
- *Extreme Earth: Waterfalls* by Patricia Corrigan

CHAPTER 17
- *One Well: The Story Of Water On Earth* by Rochelle Strauss
- *Planet Earth: Rivers and Lakes* by Rani Iyer
- *The World Around Us: Rivers* by Cecilia Minden
- *Rivers, Lakes, and Oceans (The Restless Earth)* by Gretel H. Schueller

CHAPTER 18
- *10 Things I Can Do to Help My World* by Melanie Walsh
- *Earth's Resources (Gareth Stevens Vital Science: Earth Science)* by Alfred J. Smuskiewicz
- *Where Does the Garbage Go?: Revised Edition (Let's-Read-and-Find Out About Science)* by Paul Showers and Randy Chewning
- *The Adventures of a Plastic Bottle: A Story About Recycling* by Alison Inches and Pete Whitehead
- *The Three R's: Reuse, Reduce, Recycle* by Nuria Roca and Rosa M. Curto

Microscope Information

In this activity guide, I have suggested several microscope activities. These are optional and they are best utilized with older students. For the microscope work, I done my best to include links to view the slides online, whenever possible so that purchasing a microscope is not absolutely necessary for this course. However, this course does afford a lot of opportunities for students to practice making their own slides and to become comfortable with using a microscope. I have shared the information below about purchasing and using a microscope, for your convenience.

Microscope Information

If you do not already own a microscope and you have the funds to get one, I suggest purchasing one for this course. You can purchase a good quality microscope at:
- Lab Essentials, Inc. (www.labessentials.com);
- Children's microscopes (www.childrensmicroscopes.com/022a000m.html);
- Home School Science Tools (www.hometrainingtools.com).

When purchasing a microscope, you are looking for the following things:
- ☑ A compound monocular microscope;
- ☑ A microscope with 4x, 10x, and 40x objective lenses at a minimum (NOTE – *The eyepiece should also give 10x magnification, which then will allow you to look at an object at 40x, 100x, and 400x magnification.*);
- ☑ A microscope with separate coarse and fine adjustment knobs;
- ☑ A good light source. (NOTE – *The best light source is a fluorescent bulb. Do not get one with mirror illumination.*)

If you don't know how to use a microscope, see this website for directions:

   http://www.microscope-microscope.org/basic/how-to-use-a-microscope.htm

For most of the microscope assignments, you will be making your own slides. If you don't know how to prepare a slide, check out the following post for more information on how to make your own dry and wet mount slides.

   http://elementalblogging.com/using-microscope-for-homeschool-science/

A Palm-sized Option

Many of the microscope assignments in this guide could also been done with a palm-sized microscope. You won't see quite as much as you can with a full-sized microscope, but this a much less expensive option! Here's a look at what a palm-sized microscope can do:

   http://sassafrasscience.com/palm-sized-microscope-review/

Demonstration Supplies Listed By Chapter

Chapter 1: Observing the Weather
   No supplies needed.

Chapter 2: anemometer
   5 Paper cups
   4 Straws
   6" to 8" Thin wooden dowel (about the diameter of a pencil)
   Tape
   Hole Punch
   Pencil

Chapter 3: tornado in a Bottle
   2 Soda bottles
   Duct tape
   Water

Chapter 4: Rain Gauge
   Plastic water bottle
   Duct tape
   Permanent marker
   Small marbles or rocks
   Ruler

Chapter 5: Storm in a Glass
   Clear glass jar
   Jar lid or bowl
   Ice cubes
   Warm water

Chapter 6: Snowflakes
   Glass Jar
   2 Pipe cleaners
   Pencil
   Borax (Note – This can be found in the laundry detergent aisle of the grocery store.)
   Water

Chapter 7: Expanding Ice
   Small paper cup
   Water

Chapter 8: Day or Night
   A globe (or large ball)
   A desk lamp
   A Post-it tab (or another type of removable marker)

Chapter 9: Drought Crust
   Dark construction paper (black or brown)
   Water
   Salt

Chapter 10: Barometer
   Clear plastic cup
   Soda bottle
   Blue food coloring
   Water
   Marker

Chapter 11: Cloud in a Bottle
   Hot water
   Glass jar with lid
   Crushed ice
   Match

Chapter 12: Water Cycle in a Bag
   Plastic baggie
   Warm water
   Tape

Chapter 13: Soil Test
   Soil sample
   Coffee filter
   Rubber band
   2 Cups
   Distilled water
   Aquarium test strip (one that tests the pH and nitrate levels)

Chapter 14: Moving currents
   Water
   Cup
   Ice
   Bowl
   Blue food coloring

Chapter 15: Ocean Float
   2 Eggs
   2 Tall Cups
   Water
   Salt

Chapter 16: Groundwater Filter
   Plastic bottle
   Cotton balls

Gravel
Sand
Soil
Duct tape
Water

Chapter 17: River Erosion
Flour
Aluminum pan
Eye dropper
Water

Chapter 18: Recycling Plan
Recycling bins

Project and Activity Supplies Listed by Chapter

The projects and activities listed in this guide are optional, so you may not need all of these supplies. However, this list has been provided for your convenience. If you do decide to do these projects, in addition to the items listed each week you will need clear tape, glue, scissors, a variety of paint colors, and a set of markers.

Chapter 1
No additional supplies needed.

Chapter 2
Kite
Straw
Paper
Paint
Microscope slides
Vaseline

Chapter 3
Straw
Dirt or dust
Shallow pan

Chapter 4
Clear glass
Shaving cream
Blue food coloring
Water
Paper
Microscope slides

Chapter 5
Brown paper bag
Balloon
Fluorescent light bulb

Chapter 6
Box of cornstarch
Can of shaving cream
Epsom salts
Water
Food coloring
Paper
Microscope slides

Chapter 7
Air dry clay
Brown pipe cleaners
Felt (green, red, orange, and/or yellow)
Clear glass
Crushed ice
Salt

Chapter 8
Flour
Vegetable oil
Sand
Microscope slides

Chapter 9
Materials will vary based on how the students choose to represent an oasis.

Chapter 10
White paint
Cotton balls
Blue construction paper

Chapter 11
Cotton balls
Blue construction paper

Chapter 12
Dry ice
A shallow container
Water

Chapter 13
No additional supplies needed.

Chapter 14
Plastic bottle
Water
Blue food coloring
Oil
Duct tape
Coral sample

Chapter 15
Corn syrup
Dish soap
Water
Oil
Rubbing alcohol
Black, purple, and blue food coloring
Plastic water bottle

CHAPTER 15 (continued)
Opaque liquid soap that contains glycol stearate (such as the Softsoap brand)
Duct tape

CHAPTER 16
Sponge
Bar of soap, like Ivory

CHAPTER 17
Materials will vary based on how the students choose to represent the three stages of a river.

CHAPTER 18
No additional supplies needed.

The Sassafras Guide to the Characters Found in Volume 4: Earth Science

Throughout the Book*

- ★ **Blaine Sassafras** – The male Sassafras twin, also known as Train. So far this summer, he has swung upside down in the trees, fallen out of a heliquickter, and lost his phone multiple times.
- ★ **Tracey Sassafras** – The female Sassafras twin, also known as Blaisey. So far this summer, she has been kidnapped by an Amazonian tribal leader, caught in a rockslide, and trapped inside a box.
- ★ **Cecil Sassafras** – The Sassafras twins' crazy, but talented, uncle. He is eccentric and messy, but his brilliant mind co-invented the invisible zip lines and several other contraptions.
- ★ **President Lincoln** – Uncle Cecil's lab assistant, who also happens to be a prairie dog. He doesn't say much, but his talent has been used to create amazing presentations, fix glitches, and co-invent the invisible zip lines.
- ★ **The Man with No Eyebrows** – He has no eyebrows and seems to be trying to sabotage the twins at every stop. He has broken into Cecil's lab and has been spying on Cecil's every move.

(*__Note__ – These characters also appeared in the first three volumes of *The Sassafras Science Adventures* series.)

Cecil's Neighborhood (Chapters 1 & 18)

- ★ **Old Man Grusher's Dog** – Also known as the "guardian beast". This miniature poodle loves to chase Cecil Sassafras.
- ★ **Mrs. Pascapali (paz-kah-pah-LEE)** – She is Uncle Cecil's neighbor who lives at 1106 North Pecan Street.
- ★ **Preston** – He is the squeaky and skinny teenaged clerk of the Left-handed Turtle Market.

Oklahoma City (Chapters 2 & 3)

- ★ **Sylvia Thunderstone** – The twins' local expert for their time in the Oklahoma prairie. She is a native Oklahoman and the meteorologist in charge of Lucille, the storm-chasing vehicle.
- ★ **Sylvester Hibbel (Doc)** – He is a traveling salesman, cowboy enthusiast, and inventor of several medicinal elixirs. The twins first met him during their anatomy leg.
- ★ **Jayman** – He is a friend and colleague of Dr. Thunderstone. He is responsible for relaying weather information from the main station to Lucille.

The Congo (Chapters 4 & 5)

- ★ **Carver Brighton** – The twins' local expert for their time in the Democratic Republic of the Congo. He is a professor of geochemistry who is serving as the scientific expert for the Giant Bonobo Diamond Treasure Hunt.
- ★ **Garfield T. Wellington the Fourth** – He is the benefactor of the Giant Bonobo Diamond Treasure Hunt.
- ★ **Stuart Dimsley** – He is a long-time rival of Carver Brighton's and a professor of cultural studies. He serves as the cultural expert for the Giant Bonobo Diamond Treasure Hunt.
- ★ **Bakaza (bah-KAH-zah)** – He is the Congolese guide and trailblazer for the Giant Bonobo Diamond Treasure Hunt.
- ★ **Chief Wazabanga (wah-zah-BANG-ah)** – He is the chief of the pygmy warriors that the treasure hunters run into while on their search.

Patagonia (Chapters 6 & 7)

- ★ **Hawk Talons** – He is the twins' local expert for their Patagonia leg and host of the "Out of the Office" TV show. He is an adventurer, scientist, survivalist, and member of Antarctica's Special Forces.
- ★ **Ted** – He is one of the workers at the Q.B. Cubicles office and is very reluctantly participating in the latest episode of "Out of the Office".

- ★ **Mitchell** – He is also an employee at Q.B. Cubicles that is participating in the latest episode of "Out of the Office." He is excited to be a part of the challenge, but is not the sharpest tool in the shed.
- ★ **Tammy** – She is another one of the workers from Q.B. Cubicles participating in the latest episode of "Out of the Office".
- ★ **Barbara** – She is a bit of a hypochondriac who is also from Q.B. Cubicles and is participating in the latest episode of "Out of the Office."

Mongolia (Chapters 8 & 9)

- ★ **Ganzorig (gan-ZOR-ig) Buri** – He is the twins' local expert for their time in the Mongolian Desert. Ganzorig is a college student returning home for his summer break.
- ★ **Solongo (so-LONG-o)** – A close friend of Ganzorig's. She is from the same Buri village as he is.
- ★ **Avargatom (AH-var-gat-um)** – A band of large-statured raiders who travel throughout the Mongolian desert stealing from small villages.
- ★ **Dariin (dar-EEN)** – One of Ganzorig's brothers.
- ★ **Khulan (KOO-lan)** – One of Ganzorig's brothers.

Pakistan (Chapter 10 & 11)

- ★ **Atif (A-teef) Jilani (JEE-lahn-nee)** – He is the quiet, steady, and knowledgeable local expert for the twins' time in Pakistan. He is known to everyone as the Shepherd.
- ★ **Javeria (ha-ver-EE-ah)** – She is an orphan and one of the Shepherd's apprentices.
- ★ **Aazmi (ahz-MEE)** – He is an orphan and one of the Shepherd's apprentices.
- ★ **Tariq (Tah-REEK)** – He is an orphan and one of the Shepherd's apprentices.
- ★ **Qaiser (KA-zer) Qazi (KA-zee)** – An infamous thief known as the Raider.
- ★ **Naveed (nah-VEED)** – A former student of the Shepherd's.
- ★ **The Magistrate** – A governor in the Karakoroam region of Pakistan; his word is law.

Alaska (Chapter 12 & 13)

- ★ **Summer Beach** – The loveable, excitable scientist and Alaskan local expert. Blaine and Tracey have gotten to know and love her throughout their adventure.
- ★ **Ulysses S. Grant** – Summer's lab assistant, who happens to be an arctic ground squirrel. He is the inventor of the robot squirrels.
- ★ **Yotimo** – The stoic Alaskan native who saved Tracey from a polar bear during the twins' zoology leg.
- ★ **Skeeter and Tina Romig (ROOM-ig)** – Friends of Summer's who are science teachers that recently moved to Alaska.

The Pacific Ocean (Chapters 14 & 15)

- ★ **Billfrey Battaballabingo** – The twins' local expert for their time in the Pacific Ocean. He is a marine biologist who has spent far too much time alone on the Western Garbage Patch.
- ★ **Cantankerous Carl** – Billfrey's ladder-man companion.

- ★ **Sticky Fingers Stevie** – Billfrey's coat-rack-man companion.
- ★ **Mr. and Mrs. Osodarling** – The broom and mop couple who keep Billfrey company.
- ★ **Ig** – Billfrey's mannequin companion.
- ★ **Peach Beard** – The not-too-bright captain of the P.R.O. Pirates that the twins first met on their zoology leg.

Switzerland (Chapter 16 & 17)

- ★ **Evan DeBlose** – The Triple S agent who serves as the twins' local expert for their time in Switzerland, a.k.a., Agent Pork.
- ★ **Jorgen Wuthrich** – Triple S agent and DeBlose's partner, a.k.a. Agent Beans.
- ★ **Yuroslav Bogdanovich** – He is the rogue scientist and evil villain who travels around Europe to carry out his evil schemes. The twins first met him during their botany leg.
- ★ **Adriana Archer** – Triple S agent who is goes by the name Agent Mac. She is partners with Agent Zwyssig.
- ★ **Gottfried Zwyssig (zzz-WHY-zig)** – Triple S agent who is goes by the name Agent Cheese. He is partners with Agent Archer.
- ★ **Captain Marolf** – Head of the Triple S Agency.
- ★ **Q-Tip** – The Triple S's expert in technologizing.

Chapter 1: Embarking on Earth Science

Chapter Summary

The chapter opens with Blaine, Tracey, and Uncle Cecil making the harrowing journey to the Left-Handed Turtle Market, despite the Guardian Beast, a.k.a., Old Man Grusher's poodle. Cecil forgets his wallet and so the trio has to return home. The chapter flashes over to the Man with No Eyebrows, where we learn more about his latest plan to stop the twins—the Forget-O-Nator. Back in the lab, the twins watch President Lincoln's review video from their botany adventure. The twins also learn about the new "TASER" app, before heading back to the market with their uncle. One the way back, the Guardian Beast finds the trio, a chase ensues, and Blaine is separated from the other two.

Supplies Needed

Demonstration	Projects and Activities
• No Supplies Needed	• No Additional Supplies Needed

Optional Schedules for Two-Days-A-Week

Day 1	Day 2
☐ Read Chapter 1 in *SSA* Volume 4: Earth Science*.	☐ Read the assigned pages from the encyclopedia of your choice; write narration on the Earth Science Notes Sheet on SL pg. 5.
☐ Set up your students' SCIDAT logbook.	☐ Read one of the additional living books from your library; write narration on the Earth Science Notes Sheet on SL pg. 6.
☐ Go over the vocabulary words and enter them into the ES** Glossary on SL*** pg. 95.	☐ Do the copywork or dictation assignment and add it to the ES Notes sheet on SL pg. 6.
☐ Do the demonstration entitled "Observing the Weather"; write observations on SL pg. 5.	☐ Play a game of "I Spy."

Optional Schedule for Five-Days-A-Week

Day 1	Day 2	Day 3	Day 4	Day 5
☐ Read the section entitled "Entering the territory..." of Chapter 1 in *SSA Volume 4: Earth Science*.	☐ Read the section entitled "A Photosyntastic Program" of Chapter 1 in *SSA Volume 4: Earth Science*.	☐ Read the assigned pages from the encyclopedia of your choice; write narration on the ES Notes Sheet on SL pg. 5.	☐ Read one of the additional living books from your library; write narration on the ES Notes Sheet on SL pg. 6.	☐ Do the copywork or dictation assignment and add it to the ES Notes sheet on SL pg. 6.
☐ Set up your students' SCIDAT logbook.	☐ Do the demo**** entitled "Observing the Weather"; write observations on SL pg. 5.			☐ Play a game of "I Spy".
☐ Go over the vocabulary words and enter them into the ES Glossary on SL pg. 95.				

*SSA = *The Sassafras Science Adventures*
**ES = Earth Science
***SL = *The Official Sassafras SCIDAT Logbook: Zoology Edition*
****demo = demonstration

Science-Oriented Books

Living Book Spine
- Chapter 1 of *The Sassafras Science Adventures Volume 4: Earth Science*

Optional Encyclopedia Readings
- *Basher Science Planet Earth* pp. 70-71 (Weather), pp. 72-73 (Climate)
- *Usborne Children's Encyclopedia* pg. 8 (1st half of Our Planet)
- *Discover Science Weather* pp. 6-7 (What is weather?)
- *Usborne Encyclopedia of Planet Earth* pp. 78-79 (What is weather?)

Additional Living Books
- *On the Same Day in March: A Tour of the World's Weather* by Marilyn Singer and Frane Lessac
- *Weather and Climate: Geography Facts and Experiments* (Young Discoverers Series) by Barbara Taylor

Notebooking (SCIDAT Logbook Information)

This week, you will set up the students' SCIDAT logbook. You can use blank sheets of copy paper with dividers for each section or purchase *The Official Sassafras Student SCIDAT Logbook: Earth Science Edition* with all the pages and pictures from Elemental Science. Below is an explanation of each of the student sheets.

Climate Sheets
The purpose of these sheets is to give the students an opportunity to work on their mapping skills as they study the climates different biomes around the world.

- **Area Map** – The students will color and label the area of the globe that the twins visited that has the same general climate.
- **Climate Information** – Have the students enter any of the information they learned about the biome's climate, such as average rainfall, average temperatures, and so on.
- **Interesting Facts** – Have the students enter any interesting information they have learned about the.
- **Other Types** – Have the students enter the names of any of the similar biomes they learned about. For example, when the twins are in the tropical rainforest, the students can enter "temperate rainforest" into this box. If the students are older, have them also include information about how the other type differs from the original biome's climate from the sheet.

Weather Record Sheets
The purpose of these sheets is to give the students an opportunity to work on their meteorological skills. You can have them enter the weather from your area over the two weeks they study a particular set of chapters. Alternatively, you can have the students look up and enter the weather information from the actual area that the twins have visited in the set of chapters.

- **High** – Have the students enter the high temperature for each particular day.
- **Low** – Have the students enter the low temperature for each particular day.
- **Rainfall** – Have the students enter the amount of rainfall, if any, that occurred on the particular day.
- **Conditions** – Have the students enter information about the day's weather conditions, such as

sunny or cloudy. If you have younger students, you can have them cut out the pictures for each type of weather. Templates of the weather pictures can be found in the Appendix on pg. 99.

Earth Science Record Sheets

The purpose of these sheets is for the students to record what they have learned about the various topics that are introduced in *The Sassafras Science Adventures Volume 4: Earth Science*.

Information Learned
The students should color the picture above the box, if they desire, and enter any information that they have learned about the particular topic.

Earth Science Notes Sheets

The purpose of these sheets is for the students to record any additional information that they have learned during their study of earth science. You can use these sheets to record additional narrations, copywork, or dictation assignments.

Project Record Sheets

The purpose of these sheets is for the students to record the projects they have done during the course of their study of earth science.

Earth Science Glossary

The purpose of the glossary is for the students to create a dictionary of terms that they have encountered while reading *The Sassafras Science Adventures Volume 4: Earth Science*. They can look up each term in a science encyclopedia or in the glossary included on pp. 109-110 of this guide. Then, have the students copy each definition onto a blank index card or into their SCIDAT logbook. They should also illustrate each of the vocabulary words. (**NOTE** – *In The Official Sassafras Student SCIDAT Logbook: Earth Science Edition, these pictures are already provided.*) This week, have the students look up the following terms:
- **WEATHER** – Conditions, like windy, cloudy, sunny, or rainy, that change daily.
- **CLIMATE** – The average weather in an area over a given period of time.

For each of these sheets, you can have the students enter information only from *The Sassafras Science Adventures Volume 4: Earth Science*, or you can have them do additional research to gather more facts. What you choose to do will depend on the ages and abilities of your students.

Scientific Demonstration: Observing the Weather

Begin by taking a moment to discuss the difference types of weather you can have in your area (i.e., sunny, windy, hot, cold, rainy, and so on). You can also discuss how important observation skills are for the scientist who is studying the weather. You can view the following blog posts for more information on the subject.
- http://elementalscience.com/blogs/news/63858627-observation-is-key
- http://elementalscience.com/blogs/homeschool-science-tips/71117699-3-ways-to-work-on-observation

Explain that, today, the students are going to practice their observation skills while finding out what type of weather you can find in your area. Then, take a walk in your neighborhood or on a nearby nature trail. Allow the students to make observations and ask questions. Ask the students:
⇒ *What is the weather like today?*
⇒ *What is the weather usually like in the different seasons?*

Allow the students to observe the environment and find clues from there. You can record their answers on the sheet provided in the SCIDAT Logbook.

Multi-Week Projects and Activities

Multi-week Projects

✂ **Weather Poster** – Over the coming weeks, you can have the students create a poster for each month's weather over your journey through earth science. You can download a current month-at-a-glance style calendar from here:

🖱 http://www.donnayoung.org/calendars/vertical-monthly-calendar.htm#block

Then, have the students determine the overall weather for each school day. You can use the weather pictures in the Appendix on pg. 99 for the daily weather, or have the students draw their own pictures. This week, create your first calendar and add a weather picture for each day.

Activities For This Week

✂ **I Spy** – Play a game of "I Spy" to help the students work on their observation skills.

✂ **Climates** – Have the students research the climate in which they live. Then, have them write a few sentences or draw a pictures to represent what they have learned. If you have them each write a brief paragraph, their reports could include the average yearly rainfall, the typical weather patterns, and average monthly temperatures for the year.

Memorization

Copywork/Dictation

☞ **Copywork Selection**

The weather can be windy, sunny, rainy, hot, or cold. It changes each day.

☞ **Dictation Passage**

Climate describes the average weather in an area over a given period of time. Weather refers to the exact daily conditions, such as windy, cloudy, or rainy, in an area.

Notes

Chapter 2: O-o-o-o-klahoma

Chapter Summary

The chapter opens with Cecil and Tracey making it back to the market, where they pay for the groceries. Tracey takes off on the zip lines to their first earth science location, hoping that she will find Blaine there. Instead, Tracey finds her old local expert, Doc Hibbel, and meets her new expert, Sylvia Thunderstone, as she learns about wind and Lucille the storm-chasing vehicle. We then flash over to Blaine, who is waking up in the Man with No Eyebrows's basement. The MWNE puts him in the Forget-O-Nator, planning to erase his memory. Blaine thinks quickly and uses his phone to tase the inside of the machine. Meanwhile, Tracey learns the meaning behind Sylvia's last name and a bit more about global wind patterns. The chapter ends as the tornado sirens sound in the Cowboy Hall of Fame.

Supplies Needed

Demonstration	Projects and Activities
• 5 Paper cups, 4 Straws, 6" to 8" Thin wooden dowel, Tape, Hole punch, Pencil	• Kite, Straw, Paper, Paint • Microscope slides, Vaseline

Optional Schedule for Two-Days-a-Week

Day 1	Day 2
☐ Read the section entitled "Where the wind..." of Chapter 2 in *SSA Volume 4: Earth Science*.	☐ Read the section entitled "Easterlies, Westerlies..." of Chapter 2 in *SSA Volume 4: Earth Science*.
☐ Fill out an Earth Science Record Sheet on SL pg. 9 on wind.	☐ Fill out an Earth Science Record Sheet on SL pg. 10 on global wind patterns.
☐ Add weather to the Weather Information Sheet on SL pg. 8.	☐ Add facts to the Climate Information on SL pg. 7; Add weather to the Weather Information Sheet on SL pg. 8.
☐ Go over the vocabulary words and enter them into the Earth Science Glossary on SL pg. 95.	☐ Do the demo entitled "Anemometer"; write information learned on SL pg. 13.
☐ Do the copywork or dictation assignment and add it to the Earth Science Notes on SL pg. 13.	☐ Work on one or all of the multi-week activities.

Optional Schedule for Five-Days-a-Week

Day 1	Day 2	Day 3	Day 4	Day 5
☐ Read the section entitled "Where the wind..." of Chapter 2 in *SSA Volume 4: Earth Science*.	☐ Read the section entitled "Easterlies, Westerlies..." of Chapter 2 in *SSA Volume 4: Earth Science*.	☐ Read one or all of the assigned pages from the encyclopedia of your choice; write narration on the Earth Science Notes Sheet on SL pg. 13.	☐ Read one of the additional library books.	☐ Do the copywork or dictation assignment and add it to the Earth Science Notes sheet on SL pg. 13.
☐ Fill out an Earth Science Record Sheet on SL pg. 9 on wind.	☐ Fill out an Earth Science Record Sheet on SL pg. 10 on global wind patterns.	☐ Do the demo entitled "Anemometer"; write information learned on SL pg. 13.	☐ Go over the vocabulary words and enter them into the Earth Science Glossary on SL pg. 95.	☐ Add weather to the Weather Information Sheet on SL pg. 8.
☐ Add weather to the Weather Information Sheet on SL pg. 8.	☐ Add facts to the Climate Information on SL pg. 7.		☐ Choose one of the activities for the week to do; fill out the project record sheet on pg. 15.	☐ Work on one or all of the multi-week activities.

Science-Oriented Books

Living Book Spine
- Chapter 2 of *The Sassafras Science Adventures Volume 4: Earth Science*

Optional Encyclopedia Readings
- *Basher Science Planet Earth* pp. 86-87 (Wind)
- *Usborne Children's Encyclopedia* pg. 15 (Section on Windy Weather)
- *Discover Science Weather* pp. 16-17 (Blowing Around)
- *Usborne Encyclopedia of Planet Earth* pg. 84 (Windstorms – Intro and Coriolis effect)

Additional Living Books
- *Wind* by Marion Dane Bauer and John Wallace
- *Feel the Wind* (Let's-Read-and-Find... Science 2) by Arthur Dorros
- *The Wind Blew* by Pat Hutchins
- *Like a Windy Day* by Frank Asch

Notebooking (SCIDAT Logbook Information)

This week, you can have the students begin to fill out the Climate Sheet for the Oklahoman Prairie. They can also fill out the first part of their weather record sheets and the logbook sheets for wind and global wind patterns. Here is the information they could include:

Climate Sheets
Area Map
Have the students color the region where the Great Plains prairie is found. Have the students put a star on Oklahoma City. (*See map for answers.*)

Climate Information
This week, the students could include the following:
- ⇨ *The average temperature on the prairie can be as low as -20°F in the winter and as high as 100°F in the summer, but the average is around 20°F in January and around 70°F in July.*
- ⇨ *The average rainfall is between ten and thirty inches, but most of that occurs in the summer months.*

Interesting Facts – Answers will vary.

Other Types
This week, the students could include the following:
- ⇨ *There are two types of grasslands: temperate and tropical.*
- ⇨ *The savannah, which is a tropical grassland, for instance, has a hot wet season that lasts for a few months and a slightly cooler dry season that lasts for about eight months.*
- ⇨ *The temperate prairie has cold winters and warm summers, just like the steppes of Europe and the pampas of South America.*

Weather Record Sheets
Have the students record the weather from your area or from Oklahoma City over the week.

Earth Science Record Sheets
Wind
Information Learned
- ⇨ *Wind is the movement of atmospheric gases on a large scale.*

⇨ *Wind is the movement of air.*
⇨ *We describe wind using two factors – speed and direction.*
⇨ *Wind is caused by the uneven heating of the surface of the Earth. The surface is a mixture of land and water that each absorb heat from the sun's rays in differing amounts.*
⇨ *During the day, the sun heats up the surface of the Earth and the air around it. The air over land heats up faster than the air over the water. Also, the air over places that receive direct sunlight heats up faster than the air over land that receives indirect sunlight. Since the warm air weighs less, it rises, a change in air pressure occurs, and the cool air moves in to replace the space where the warm air was. This movement of warm and cool air causes wind.*
⇨ *At night, the air over land cools more quickly than the air over water, so wind is created once more.*
⇨ *When there is lots of wind, we can harness the power of the wind and turn it into energy that we can use. This is known as wind power. Today, we use wind power to generate electricity, but in the past, it was used to pump water on the prairies.*

Global Wind Patterns
Information Learned

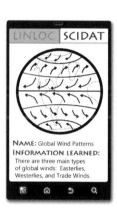

⇨ *The movement of air around the globe is known as the global wind patterns. On a large scale, the winds that circle the Earth are created because the land at the equator is heated more than the land at the poles.*
⇨ *Another factor that affects the winds around the globe is the spinning motion of the earth. This is known as the Coriolis Effect.*
⇨ *There are three main types of global winds – the easterlies, westerlies, and trade winds*
   1. *Trade winds – These winds are found near the equator. They flow north or south towards the equator and curve west due to the spin of the Earth.*
   2. *Prevailing westerlies – These winds are found in between the equator and the poles. They blow slightly towards the poles, from the west to the east.*
   3. *Polar easterlies – These winds are found near the north and south poles. They blow up to the poles and curve from east to west.*
⇨ *A jet stream is a river of fast-moving air about five to nine miles above the Earth's surface. They form at the boundaries where the polar and temperate or tropical air meet. Because of the effect of the rotation of the Earth, the jet streams flow from west to east in a wave-like manner.*

Vocabulary
Have the older students look up the following terms in the glossary in the Appendix on pp. 109-110 or in a science encyclopedia. Then, have them copy each definition onto a blank index card or into their SCIDAT logbook.
   ⇨ WIND – The movement of air in the atmosphere created by temperature differences.
   ⇨ GUST – A short burst of wind moving at a high speed.

Scientific Demonstration: Anemometer
Materials
☑ 5 Paper cups
☑ 4 Straws
☑ 6" to 8" Thin wooden dowel (about the diameter of a pencil)
☑ Tape

- ☑ Hole punch
- ☑ Pencil

Procedure
1. Have the students punch a single hole in the side of each of four cups, about halfway down from the rim. Then, on the fifth cup, have them punch two sets of holes directly across from each, other about half an inch down from the rim. The holes on the fifth cup should line up to create a cross through the middle of the cup. Finally, use a pencil to poke a hole in center of the bottom of the fifth cup.
2. Next, insert a straw into the four cups with the single hole and secure it in place. Then, insert the four straws in the holes on the side of the fifth cup so that the cups are tilted sideways and the four straws meet in the center of the fifth cup. Use the tape to secure the four straws in an "x". (*See the diagram for a visual reference.*)
3. Now, push the dowel rod into hole in the bottom of the fifth cup. The students have now created a rudimentary anemometer.
4. Head outside to test their creation. Once outside, place the anemometer in dirt or hold it in your hand. Their device should stand upright, but still be free to turn in the wind.

Explanation
The students should see that when the wind blows, their anemometer turns. They should also observe that the faster the wind blows, the quicker the device turns. An anemometer is designed to measure wind speed and it a common instrument found at a weather station.

Take it further
Have the students make a simple wind vane, which is a device to measure wind direction. A light ribbon or streamer will work for this. Simply have them hold one end of the ribbon in one hand and hold that hand above their heads. They can observe which way the ribbon moves in the breeze to determine which direction the wind is blowing.

Multi-Week Projects and Activities

Multi-week Projects
✂ **Weather Poster** – Have the students add to their weather poster this week. (*Weather template pictures can be found in the Appendix on pg. 8.*)

Activities For This Week
✂ **Fly a Kite** – If you have an exceptionally windy day, have the students head outside to fly a kite.

✂ **Wind Patterns** – Have the students draw and label the three main wind patterns on the globe. You can use the map template found in the Appendix on pg. 106. See the accompanying image for the answers.

✂ **Wind Painting** – You will need a straw, paper, and paint for each student. Have the students place a drop of paint at one end of their paper. Then, have them use the straw to blow the paint into a design!

✂ **Microscope Work** – Have the students look at what is found in the wind under the microscope. They can do this by covering a slide with a thin layer of Vaseline before setting the slide out on a windy day. Prop the slide up on

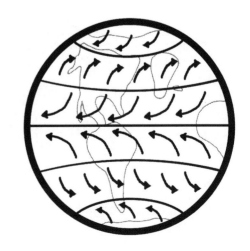

a chair, so that it will be in the path of the wind. Wait about ten minutes before bringing the slide inside to view it under the microscope. Have the students complete one of the microscope worksheets found on pp. 97-98 of the Appendix.

Memorization

Copywork/Dictation

☞ **Copywork Sentence**
Wind is the movement of air. We describe wind using two factors – speed and direction.

☞ **Dictation Selection**
Wind is the movement of atmospheric gases on a large scale. It is caused by the uneven heating of the surface of the Earth. We describe wind using two factors – speed and direction. The movement of air around the globe is known as the global wind patterns.

Notes

Chapter 3: Lucille's First Rodeo

Chapter Summary

The chapter opens with Cecil finding Blaine wandering around outside. He gets they boy back to the lab and then sends him off to Oklahoma on the zip lines. In the meantime, we find out that Blaine's taser-app destroyed the Forgot-O-Nator, which the Man with No Eyebrows vows to rebuild. Back in Oklahoma, Tracey finds Blaine and introduces him to Sylvia, who teaches them about downbursts before they hop into Lucille to chase down a tornado. As they move into position, Sylvia shares about tornadoes with the twins. They anchor the vehicle down in a field directly in the path of the storm. As the tornado rapidly approaches, Sylvia notices some children trapped in a nearby building.

Supplies Needed

Demonstration	Projects and Activities
• 2 Soda bottles, Duct tape, Water	• Straw, Dirt or dust, Shallow pan

Optional Schedule for Two-Days-A-Week

Day 1	Day 2
☐ Read the section entitled "Damaging Downbursts" of Chapter 3 in *SSA Volume 4: Earth Science*.	☐ Read the section entitled "Thunderstone's Tornado" of Chapter 3 in *SSA Volume 4: Earth Science*.
☐ Fill out an Earth Science Record Sheet on SL pg. 11 on downbursts.	☐ Fill out an Earth Science Record Sheet on SL pg. 12 on tornadoes.
☐ Add facts to the Climate Information on SL pg. 7; Add weather to the Weather Information Sheet on SL pg. 8.	☐ Add weather to the Weather Information Sheet on SL pg. 8; Go over the vocabulary word and enter it into the Earth Science Glossary on SL pg. 96.
☐ Do the copywork or dictation assignment and add it to the Earth Science Notes on SL pg. 14.	☐ Do the demo entitled "Tornado in a Bottle"; write information learned on SL pg. 14.
	☐ Work on one or all of the multi-week activities.

Optional Schedule for Five-Days-A-Week

Day 1	Day 2	Day 3	Day 4	Day 5
☐ Read the section entitled "Damaging Downbursts" of Chapter 3 in *SSA Volume 4: Earth Science*.	☐ Read the section entitled "Thunderstone's Tornado" of Chapter 3 in *SSA Volume 4: Earth Science*.	☐ Read one or all of the assigned pages from the encyclopedia of your choice; write narration on the Earth Science Notes Sheet on SL pg. 14.	☐ Read one of the additional library books.	☐ Do the copywork or dictation assignment and add it to the Earth Science Notes sheet on SL pg. 14.
☐ Fill out an Earth Science Record Sheet on SL pg. 11 on downbursts.	☐ Fill out an Earth Science Record Sheet on SL pg. 12 on tornadoes.	☐ Do the demo entitled "Tornado in a Bottle"; write information learned on SL pg. 14.	☐ Go over the vocabulary word and enter it into the Earth Science Glossary on SL pg. 96.	☐ Add weather to the Weather Information Sheet on SL pg. 8.
☐ Add facts to the Climate Information on SL pg. 7.	☐ Add weather to the Weather Information Sheet on SL pg. 8.		☐ Choose one of the activities for the week to do; fill out the project record sheet on pg. 16.	☐ Work on one or all of the multi-week activities.

Science-Oriented Books

Living Book Spine
- 📖 Chapter 3 of *The Sassafras Science Adventures Volume 4: Earth Science*

Optional Encyclopedia Readings
- *Basher Science Planet Earth* pp. 110-111 (Grassland)
- *Discover Science Weather* pp. 18-19 (Wild winds)
- *Usborne Children's Encyclopedia* pg. 16 (Types of Storms)
- *Usborne Encyclopedia of Planet Earth* pg. 85 (Tornadoes, Tornado alley, and Sea spouts)

Additional Living Books
- 📖 *Tornadoes!* by Gail Gibbons
- 📖 *Tornadoes* by Seymour Simon
- 📖 *Tornado Alert (Let's-Read-and-Find-Out Science 2)* by Franklyn M. Branley and Giulio Maestro
- 📖 *A Grassland Habitat (Introducing Habitats)* by Kelley Macaulay and Bobbie Kalman
- 📖 *Grasslands (About Habitats)* by Cathryn P. Sill

Notebooking (SCIDAT Logbook Information)

This week, you can have the students add to the Climate Sheet for the Oklahoman Prairie. They can also fill out the second part of their weather record sheets and the logbook sheets for downbursts and tornadoes. Here is the information they could include:

Climate Sheets
Climate Information – This week, the students could include the following:
- ⇨ The warm, humid summer on the grassland allows the grass to grow very tall, but it also sparks lots of storms.
- ⇨ On the grasslands, there are no natural barriers like trees or mountains, so there is a lot of wind.

Interesting Facts – Answers will vary.

Weather Record Sheets
Have the students record the weather from your area or from Oklahoma City over the week.

Earth Science Record Sheets
Downbursts
Information Learned

- ⇨ Derechos are large clusters of strong thunderstorms that form in a long line and can cause widespread wind damage. This damage is caused by the abundance of downburst winds that derecho storms can produce. These storms typically form in the late spring/summer.
- ⇨ A downburst is strong downward current of air often associated with a thunderstorm.
- ⇨ Downbursts are caused by rain-cooled air that sinks and rushes back to the ground. When it reaches ground-level, it quickly spreads out in all directions and produces strong, damaging winds.
- ⇨ Unlike tornadoes, winds in a downburst blow outwards from the point at which the wind hits the land.

This forces things out of the wind's path, instead sucking them in like a tornado would.
⇨ *These storms typically form in the late spring/summer.*

TORNADOES
INFORMATION LEARNED

⇨ *A tornado is a rapidly spinning funnel of air that touches the ground and is connected to the clouds above.*
⇨ *Before a tornado touches the ground, it is known as a funnel cloud.*
⇨ *Most tornadoes only last a few minutes, but in that time they can tear up trees and houses, plus move cars, animals, and people.*
⇨ *Tornadoes typically form from strong thunderstorms. In the midst of these storms clouds, there is hot, humid, fast-moving air moving upward and cold, dry air moving downward. These two currents spiral and spin around each other, forming a funnel. If the currents are strong enough, this funnel will reach the ground to form a tornado.*
⇨ *The Fujitsu scale is used to describe the strength of a tornado. It ranges from F0 to F5, with F5 being the strongest tornado. Each category has a wind speed range and a description of possible damage.*
⇨ *The majority of tornadoes spins at around one hundred miles per hour, which is an F1 on the Fujitsu scale. An F1 tornado can cause moderate damage, like snapping trees, blowing around mobile homes, and damaging roofs.*
⇨ *Most tornados occur in an area known as Tornado Alley, which includes the Great Plains states in the US. Over 500 tornadoes touch down in this area each ear.*

VOCABULARY

Have the older students look up the following term in the glossary in the Appendix on pp. 109-110 or in a science encyclopedia. Then, have them copy the definition onto a blank index card or into their SCIDAT logbook.

↳ TORNADO – A spinning funnel of wind that touches the ground and is also connected to the clouds above.

SCIENTIFIC DEMONSTRATION: TORNADO IN A BOTTLE
MATERIALS
☑ 2 Soda bottles
☑ Duct tape
☑ Water

PROCEDURE
1. Have the students fill one of the soda bottles two-thirds of the way with room temperature water.
2. Then, have them invert the second bottle and use the duct tape to attach the two openings together so that no water will leak out.
3. Once the two bottles are securely attached, have the students flip the bottles over and observe what happens

EXPLANATION
The students should see a vortex or funnel of form as the water moves from one bottle to the other. The swirling motion and movement of the water mimics the same conditions in the air that form a tornado.

TAKE IT FURTHER
Have the students add a bit of glitter and small sequins to the water to act as debris. Then, have the repeat the demonstration to see how debris acts in a tornado.

Multi-Week Projects and Activities

Multi-week Projects
- **Weather Poster** – Have the students add to their weather poster this week. (*Weather template pictures can be found in the Appendix on pg. 8.*)

Activities For This Week
- **Downbursts** – You will need a straw, a bit of dirt or dust, and a shallow pan. Have the students form a pile of dirt in the center of the shallow pan. Then, have them hold the straw an inch above the pile and blow a burst of air directly down on the pile, just like a downburst. Have the students observe how the dirt was moved around by the wind and the damage the blowing did to the pile.
- **Tornado Video** – Have the students watch the following National Geographic video on tornadoes:
    - https://www.youtube.com/watch?v=pSajNLBH7cA
- **Watch or Warning** – Have the students research and learn about the differences between a tornado watch and warning. (*A watch means that conditions are favorable for the weather event to occur, while a warning means that the occurrence of the given weather event is imminent.*) After they understand the differences, have the students create a poster sharing what a watch and a warning are, along with what a person should do when one or the other is issued by their local weather service.

Memorization

Copywork/Dictation
- **Copywork Sentence**
    A tornado is a spinning funnel of air that touches the ground and the clouds above.
- **Dictation Selection**
    The Fujitsu scale is used to describe the strength of a tornado. It ranges from F0 to F5, with F5 being the strongest tornado. Each category has a wind speed range and a description of possible damage. The majority of tornadoes spins at around one hundred miles per hour, which is an F1 on the Fujitsu scale.

Notes

Chapter 4: The Congolese Jungle Treasure Hunt

Chapter Summary

The chapter opens with Sylvia, the twins' local expert in Oklahoma, leaving the twins in Lucille as the tornado bears down on them. Blaine and Tracey survive the damaging winds and get back to solid ground, with a little help from Doc Hibbel. After finding Sylvia and the children she rescued, the twins zip off to their next location in the Congo. They meet a treasure hunting party led by Garfield T. Wellington the Fourth. They learn about rain and meet the other members of the party, Stuart Dimsley, Bakaza, and Carver Brighton, their local expert. The twins join the hunt for the Giant Bonobo Diamond. On the way, they learn about monsoons, and they experience the results of heavy rains firsthand as they slide through the jungle! The group ends up in a river and swims to the bank, where they encounter a band of pygmy warriors!

Supplies Needed

Demonstration	Projects and Activities
• Plastic water bottle, Duct tape, Permanent marker, Small marbles or rocks, Ruler	• Clear glass, Shaving cream, Blue food coloring, Water • Paper, Markers, Microscope slide

Optional Schedule for Two-Days-A-Week

Day 1	Day 2
☐ Read the section entitled "Carver's Rain Takes…" of Chapter 4 in *SSA Volume 4: Earth Science*. ☐ Fill out an Earth Science Record Sheet on SL pg. 19 on rain. ☐ Add weather to the Weather Information Sheet on SL pg. 18. ☐ Do the demo entitled "Rain Gauge"; write information learned on SL pg. 23. ☐ Do the copywork or dictation assignment and add it to the Earth Science Notes on SL pg. 23.	☐ Read the section entitled "Moving Monsoons" of Chapter 4 in *SSA Volume 4: Earth Science*. ☐ Fill out an Earth Science Record Sheet on SL pg. 20 on monsoons. ☐ Add facts to the Climate Information on SL pg. 17; Add weather to the Weather Information Sheet on SL pg. 18. ☐ Go over the vocabulary words and enter them into the Earth Science Glossary on SL pg. 96; Check on the rain gauge and add the data to SL pg. 25. ☐ Work on one or all of the multi-week activities.

Optional Schedule for Five-Days-A-Week

Day 1	Day 2	Day 3	Day 4	Day 5
☐ Read the section entitled "Carver's Rain Takes…" of Chapter 4 in *SSA Volume 4: Earth Science*. ☐ Fill out an Earth Science Record Sheet on SL pg. 19 on rain. ☐ Add weather to the Weather Information Sheet on SL pg. 18.	☐ Read one or all of the assigned pages from the encyclopedia of your choice; write narration on the Earth Science Notes Sheet on SL pg. 23. ☐ Do the demo entitled "Rain Gauge"; write information learned on SL pg. 23.	☐ Read the section entitled "Moving Monsoons" of Chapter 4 in *SSA Volume 4: Earth Science*. ☐ Fill out an Earth Science Record Sheet on SL pg. 20 on monsoons. ☐ Add facts to the Climate Information on SL pg. 17.	☐ Read one of the additional library books. ☐ Go over the vocabulary words and enter them into the Earth Science Glossary on SL pg. 96. ☐ Choose one of the activities for the week to do; fill out the project record sheet on pg. 27.	☐ Do the copywork or dictation assignment and add it to the Earth Science Notes sheet on SL pg. 25. ☐ Check on the rain gauge and add the data to SL pg. 25; Add weather to the Weather Information Sheet on SL pg. 18. ☐ Work on one or all of the multi-week activities.

Science-Oriented Books

Living Book Spine
- Chapter 4 of *The Sassafras Science Adventures Volume 4: Earth Science*

Optional Encyclopedia Readings
- *Basher Science Planet Earth* pp. 34-35 (Tropics), pp. 82-83 (Precipitation)
- *Usborne Children's Encyclopedia* pg. 17 (Section on Monsoon Floods)
- *Discover Science Weather* pp. 24-25 (Out in the rain), pp. 38-39 (Rain or Shine)
- *Usborne Encyclopedia of Planet Earth* pp. 62-63 (Monsoons)

Additional Living Books
- *Down Comes the Rain* (Let's-Read-And-Find... Science 2) by Franklyn Mansfield Branley
- *The Rain Came Down* by David Shannon
- *Rain* (Weather Series) by Marion Dane Bauer and John Wallace
- *A Rainforest Habitat* (Introducing Habitats) by Molly Aloian

Notebooking (SCIDAT Logbook Information)

This week, you can have the students begin to fill out the Climate Sheet for the Congolese Rainforest. They can also fill out the first part of their weather record sheets and the logbook sheets for rain and monsoons. Here is the information they could include:

Climate Sheets
Area Map – Have the students color the region where the Congolese Rainforest is found. (*See map for answers.*)
Interesting Facts – Answers will vary.
Other Types – This week, the students could include the following:
  ⇨ *Two types of rainforests—tropical and temperate.*
  ⇨ *Temperate forests get around one hundred inches of rain per year.*
  ⇨ *Tropical forests are warm and humid all the time (like the Congo), while temperate forests can have a cool season and sometimes even experience frost.*

Weather Record Sheets
Have the students record the weather from your area or from the Congo over the week.

Earth Science Record Sheets
Rain
Information Learned
  ⇨ *Rain is water falling from clouds in the form of water droplets. Water can also fall from clouds as hail, snow, or sleet.*
  ⇨ *Rain forms when warm, moist air rises and condenses to form a cloud of water vapor. The micro-droplets collect together to form bigger droplets, which fall to the ground because of gravity.*
  ⇨ *Raindrops are one-hundreth to one-tenth of an inch in diameter, so they are quite*

tiny. Very fine drops of rain fall at a rate of about two miles per hour, while very heavy drops fall as fast as eighteen miles per hour.
- ⇨ *Heavy rain can cause many problems, such as flooding and landslides.*
- ⇨ *Acid rain is rain with a low pH due to the presence of sulfur dioxide and nitrogen oxides than have been released by factories. Acid rain can damage plant life and buildings.*
- ⇨ *The highest rainfall ever recorded was in India, where they had around a thousand inches of rain fall in one year.*

Monsoons
Information Learned

- ⇨ *Monsoons are the seasonal changes in the strongest winds of a region.*
- ⇨ *The largest monsoon winds are found in Asia, but there are smaller ones in northern Australia, along the equator in Africa, and in the southwestern US.*
- ⇨ *Monsoons are the cause of the wet and dry seasons found throughout the tropics. These winds always blow from cold to warm regions.*
- ⇨ *The summer monsoon winds typically blow from the southwest regions across the ocean and onto land. The air brought onto land is rich in moisture, or humid. This causes heavy, or torrential, rainfall. The abundance of rain fills wells and aquifers that the people in the region will rely on during the dry season.*
- ⇨ *The winter monsoon winds are opposite. The wind typically blows in from the northeast regions bringing typically drier air. The exception to this is in the Pacific Northwest, where the winter monsoon winds bring about their wet season.*
- ⇨ *The Congo is affected by the Asian-Australian monsoon winds, which stretch from the northern coast of Australia up Pacific and over across the Indian Ocean to the coast of Africa.*

Vocabulary

Have the older students look up the following terms in the glossary in the Appendix on pp. 109-110 or in a science encyclopedia. Then, have them copy each definition onto a blank index card or into their SCIDAT logbook.
- ⟁ MONSOON – A season of strong winds and heavy rain.
- ⟁ PRECIPITATION – Rain, snow, sleet, or hail that falls to the ground.

Scientific Demonstration: Rain Gauge
Materials
- ☑ Plastic water bottle
- ☑ Duct tape
- ☑ Permanent marker
- ☑ Small marbles or rocks
- ☑ Ruler

Procedure
1. Begin by cutting the top off of the plastic bottle at the point where the diameter of the bottle starts to decrease. Keep the top portion of the bottle that was cut off and cover all the cut edges with duct tape to hide the sharp edges.
2. On the bottom section of the bottle, have the students use a ruler to make a scale of horizontal lines in centimeters with the permanent marker. They need to make the first line four centimeters from the bottom. Then, have them add a layer of marbles or rocks in the bottom portion of the bottle to steady the base.
3. Next, have the students pour water into the bottle up to the first line of their scale. Finally, have them

turn the top portion of the bottle upside down and place it inside the opening of the bottom portion to form a funnel-shaped lid.
4. Now the students are ready to begin measuring the rain with their bottle. The optional rain gauge chart on SL pp. 25-26 has both a predicted rain amount and an actual rain amount. If you choose to have the students predict the rainfall amount, have them listen to the weather report and then predict how much they think it is going to rain each day. You can choose to have them record the rain for 5 or 7 days a week. Either way, they will be recording their observations in a similar fashion over the next few weeks.

Explanation
The purpose of this demonstration is to allow the students to see and record the amount of rainfall found in their area.

Multi-Week Projects and Activities

Multi-week Projects
✂ **Weather Poster** – Have the students add to their weather poster this week. (*Weather template pictures can be found in the Appendix on pg. 8.*)

Activities For This Week
✂ **Indoor Rain** – Create an indoor rainstorm with the students! You will need a clear glass, shaving cream, blue food coloring, and warm water. Use the directions from the following post:
  🖱 http://sassafrasscience.com/how-to-make-indoor-rainstorm/

✂ **Raindrop Paintings** – Have the students color a sheet of paper with markers in a design of their choosing. Let their creation dry. Then, have the students dip their fingers in a bowl of water and sprinkle rain down on their artwork until they are satisfied with the results.

✂ **Acid Rain** – Have the students earn about what acid rain is and how it affects the environment. The following video is a good option for this:
  🖱 https://www.youtube.com/watch?v=Nf8cuvl62Vc

If your students are older, you can have them do a bit of additional research on the topic before they write a short, one page report on acid rain and its effects.

✂ **Microscope Work** – Have the students look at raindrops under the microscope. They can do this by setting the slide out on a rainy day. Once it is mostly covered with raindrops, bring it inside to view the slide under the microscope. Have the students complete one of the microscope worksheets found on pp. 97-98 of the Appendix. If you do not own a microscope, you can view the following image of rainwater after a volcanic eruption at the following website:
  🖱 http://all-geo.org/volcan01010/2011/04/eyjafjallajokull-anniversary/

Memorization

Copywork/Dictation
☞ **Copywork Sentence**
  Rain is water falling from clouds in the form of water droplets.

☞ **Dictation Selection**
  Rain is water falling from clouds in the form of water droplets. Rain forms when warm, moist air rises and condenses to form a cloud of water vapor. The micro-droplets collect together to form bigger droplets, which fall to the ground because of gravity. These droplets can fall as rain, hail, snow, or sleet.

Notes

Chapter 5: The Search for the Giant Bonobo Diamond

Chapter Summary

The chapter opens with the treasurer hunters in a standoff with the pygmy warriors. The warriors break the silence first and end up inviting the group back to their village. Over a meal of MVM (Machete Vegetable Medley), the Carver shares about thunderstorms. The group leaves the village and sets off to find the valley they are seeking. The group quickly finds the temple of the lost Giant Bonobo Diamond, where they find a cryptic poem and a quickly approaching flood. The flood actually turns out to help the treasurer hunters move through the temple. After a series of trials, which they overcome with the help of the poem from the entrance, the group arrives at end of a passageway where they hope to find the missing diamond!

Supplies Needed

Demonstration	Projects and Activities
• Clear glass jar, Jar lid or bowl • Ice cubes, Warm water	• Brown paper bag • Balloon, Fluorescent light bulb

Optional Schedule for Two-Days-A-Week

Day 1	Day 2
☐ Read the section entitled "Thundering Pygmy..." of Chapter 5 in *SSA Volume 4: Earth Science*.	☐ Read the section entitled "Flooded Findings" of Chapter 5 in *SSA Volume 4: Earth Science*.
☐ Fill out an Earth Science Record Sheet on SL pg. 21 on thunderstorms.	☐ Fill out an Earth Science Record Sheet on SL pg. 22 on floods.
☐ Add facts to the Climate Information on SL pg. 17; Add weather to the Weather Information Sheet on SL pg. 18.	☐ Add weather to the Weather Information Sheet on SL pg. 18; Check on the rain gauge and add the data to SL pg. 25.
☐ Do the demo entitled "Storm in a Jar"; write information learned on SL pg. 24.	☐ Go over the vocabulary words and enter them into the Earth Science Glossary on SL pp. 96-97.
☐ Do the copywork or dictation assignment and add it to the Earth Science Notes on SL pg. 24.	☐ Work on one or all of the multi-week activities.

Optional Schedule for Five-Days-A-Week

Day 1	Day 2	Day 3	Day 4	Day 5
☐ Read the section entitled "Thundering Pygmy..." of Chapter 5 in *SSA Volume 4: Earth Science*. ☐ Fill out an Earth Science Record Sheet on SL pg. 21 on thunderstorms. ☐ Add weather to the Weather Information Sheet on SL pg. 18.	☐ Read one or all of the assigned pages from the encyclopedia of your choice; write narration on the Earth Science Notes Sheet on SL pg. 24. ☐ Do the demo entitled "Storm in a Jar"; write information learned on SL pg. 24.	☐ Read the section entitled "Flooded Findings" of Chapter 5 in *SSA Volume 4: Earth Science*. ☐ Fill out an Earth Science Record Sheet on SL pg. 22 on floods. ☐ Add facts to the Climate Information on SL pg. 17.	☐ Read one of the additional library books. ☐ Go over the vocabulary words and enter them into the Earth Science Glossary on SL pp. 96-97. ☐ Choose one of the activities for the week to do; fill out the project record sheet on pg. 28.	☐ Do the copywork or dictation assignment and add it to the Earth Science Notes sheet on SL pg. 24. ☐ Check on the rain gauge and add the data to SL pg. 25; Add weather to the Weather Information Sheet on SL pg. 18. ☐ Work on one or all of the multi-week activities.

Science-Oriented Books

Living Book Spine
- Chapter 5 of *The Sassafras Science Adventures Volume 4: Earth Science*

Optional Encyclopedia Readings
- *Basher Science Planet Earth* pp. 91 (Flood)
- *Discover Science Weather* pp. 26-27 (Stormy Days), pp. 28-29 (Wet and Dry)
- *Usborne Children's Encyclopedia* pg. 17 (Section on Floods)
- *Usborne Encyclopedia of Planet Earth* pg. 86 (Sections on Floods), pp. 82-83 (Thunderstorms)

Additional Living Books
- *Flood (Capstone Young Readers)* by Alvaro F. Villa
- *National Geographic Readers: Storms!* by Miriam Goin
- *Wild Weather, Level 1 Extreme Reader (Extreme Readers)* by Katharine Kenah

Notebooking (SCIDAT Logbook Information)

This week, you can have the students add to the Climate Sheet for the Congolese Rainforest. They can also fill out the second part of their weather record sheets and the logbook sheets for thunderstorms and floods. Here is the information they could include:

Climate Sheets
Interesting Facts – This week, the students could include the following:
⇨ Tropical rainforests occur near the equator.
⇨ The equator is an imaginary line drawn around the earth equally distant from both poles, dividing the earth into northern and southern hemispheres.

Weather Record Sheets
Have the students record the weather from your area or from the Congo over the week.

Earth Science Record Sheets
Thunderstorm
Information Learned
⇨ A thunderstorm is a storm with thunder and lightning.
⇨ These storms can produce hail and heavy rain, but the biggest danger from a thunderstorm comes from lightning.
⇨ They can occur at any time, as long as there is moisture, unstable air, and lift in the atmosphere. However, they are most likely to occur during the spring and summer months in the afternoons and evenings.
⇨ Lightning is an electrical current produced by a thunderstorm. Lightning is electricity in the form of a bright flash across the sky. The electricity is produced as ice crystals within a thundercloud rub together and produce electrical charges. The negative charges in the cloud are attracted to the positively charged ground. When the two connect, a lightning bolt forms.
⇨ Thunder is the result of lightning. When a lightning bolt strikes, it opens up a small pocket of air. When the light is gone, the pocket collapses and creates a sound wave that we can hear. Even though the two events happen close together, we see lightning before we hear thunder because light travels faster than

sound.

Flood
Information Learned
⇨ *A flood is the result of heavy rains or lots of melting snow.*
⇨ *The rivers and lakes rise above their normal levels and the water spills out and over onto the surrounding land.*
⇨ *Floodwaters move slowly across flat lands, but speed up through canyons and valleys.*
⇨ *Floodwaters can be very dangerous. A little more than six inches of water can knock people off their feet and potentially sweep them away.*
⇨ *Floods, are one of the top weather-related killers.*
⇨ *Some floods can be predicted several days in advance. Some, known as flash floods, can come up with little to no warning.*
⇨ *A flash flood can occur after a period of intense rain, usually from a slow-moving thunderstorm. The water within the rivers and lakes rises rapidly and without warning to form a flash flood.*

Vocabulary
Have the older students look up the following terms in the glossary in the Appendix on pp. 109-110 or in a science encyclopedia. Then, have them copy each definition onto a blank index card or into their SCIDAT logbook.

- THUNDERSTORM – A storm with thunder and lightning.
- FLOOD – An overflow of a large amount of water due to heavy rains or melting snow.

Scientific Demonstration: Storm in a Jar
Materials
☑ Clear glass jar
☑ Jar lid or bowl
☑ Ice cubes
☑ Hot water

Procedure
1. Fill the jar two-thirds of the way full with hot water. (NOTE – *The jar will get very warm, so do not leave the students unsupervised at any point during this demonstration!!*)
2. Have the students quickly cover the jar with an upside down lid or bowl. Have them fill the lid and bowl with ice cubes.
3. Observe the changes that take place over the next thirty minutes.

Explanation
The students should see condensation forming after a few minutes. If they wait long enough, they will begin to see droplets form and fall in the jar. The conditions created in the jar are very similar to the ones that produce storms and rain on Earth. The warm, moist air comes in contact with cool air. Then, the water vapor condenses, collects, and falls.

Take it further
Have the students repeat the demonstration with room temperature and ice cold water to see if their results differ.

Multi-Week Projects and Activities
Multi-week Projects
✂ WEATHER POSTER – Have the students add to their weather poster this week. (*Weather template*

pictures can be found in the Appendix on pg. 8.)

ACTIVITIES FOR THIS WEEK

- **THUNDER** – You will need a brown paper bag for this activity. Have the students blow the bag up with air and twist it shut. Hold the bag at the twist with one hand and hit it hard with the other to create thunder! (*The air inside the bag is quickly compressed, forcing the bag to rip and the air to rush out. This creates a sound wave, which we can hear. This process is very similar to the one that creates thunder during a lightning storm.*)

- **LIGHTNING SPARK** – You will need a balloon and a fluorescent light bulb. Head into a darkened room. Have the students blow up the balloon and rub it on their hair or clothing, which will produce a static charge. Then, have the hold the balloon up to the light bulb and observe what happens! (*The light you see in the bulb is the result of the negative electrical charge in the balloon interacting with the positive charges in the light bulb. The same process create lightning as the electrical charges in a storm cloud interact with the positive charges on the ground.*)

- **FLOOD LEVELS** – Have the students do the following online activity to see how a rising sea level could affect where they live:
  - http://stem-works.com/subjects/7-tsunamis-floods/activities/134

MEMORIZATION

COPYWORK/DICTATION

- **COPYWORK SENTENCE**
  A flood is the result of heavy rains or lots of melting snow.

- **DICTATION SELECTION**
  Lightning is an electrical current produced by a thunderstorm. Thunder is the result of lightning. When a lightning bolt strikes, it opens up a small pocket of air. When the light is gone, the pocket collapses and creates a sound wave that we can hear.

NOTES

Chapter 6: Parachuting into Patagonia

Chapter Summary

The chapter opens with Carver and Tracey solving the mystery of where the diamond is hidden. The group finds the Giant Bonobo Diamond and the Sassafras Twins zip off to their next location—the Patagonian mountains. They land on a box at the top of a mountain in the middle of a snow storm. Blaine and Tracey quickly find out that they are not alone and that they are part of the latest episode of the reality TV show "Out of the Office." They meet the host and their local expert, Hawk Talons, who is informs the group of the first challenge, building a snow cave, and tells them about the falling snow. The twins win the first challenge and after a night in the snow, the group of Q.B. Cubicles salesmen, along with Blaine and Tracey, begin their descent down the mountain. The second "Out of the Office" challenge is to rappel down a cliff. Blaine and Tracey's quick thinking earns them another win. As the group descends, an ice storm provides the perfect time for their expert to tell them about the science behind the weather phenomenon.

Supplies Needed

Demonstration	Projects and Activities
• Glass Jar, 2 Pipe cleaners, Pencil • Borax, Water	• Box of cornstarch, Can of shaving cream • Epsom salts, Food coloring, Paper, Microscope slide

Optional Schedule for Two-Days-A-Week

Day 1	Day 2
☐ Read the section entitled "Snowy Set Downs" of Chapter 6 in *SSA Volume 4: Earth Science*.	☐ Read the section entitled "Icy Impositions" of Chapter 6 in *SSA Volume 4: Earth Science*.
☐ Fill out an Earth Science Record Sheet on SL pg. 31 on snow.	☐ Fill out an Earth Science Record Sheet on SL pg. 32 on ice storms.
☐ Add facts to the Climate Information on SL pg. 29; Add weather to the Weather Information Sheet on SL pg. 30.	☐ Add weather to the Weather Information Sheet on SL pg. 30; Check on the rain gauge and add the data to SL pg. 26.
☐ Do the demo entitled "Snowflakes"; write information learned on SL pg. 35.	☐ Go over the vocabulary word and enter it into the Earth Science Glossary on SL pg. 97.
☐ Do the copywork or dictation assignment and add it to the Earth Science Notes on SL pg. 35.	☐ Work on one or all of the multi-week activities.

Optional Schedule for Five-Days-A-Week

Day 1	Day 2	Day 3	Day 4	Day 5
☐ Read the section entitled "Snowy Set Downs" of Chapter 6 in *SSA Volume 4: Earth Science*. ☐ Fill out an Earth Science Record Sheet on SL pg. 31 on snow. ☐ Add weather to the Weather Information Sheet on SL pg. 30.	☐ Read one or all of the assigned pages from the encyclopedia of your choice; write narration on the Earth Science Notes Sheet on SL pg. 35. ☐ Do the demo entitled "Snowflakes"; write information learned on SL pg. 35.	☐ Read the section entitled "Icy Impositions" of Chapter 6 in *SSA Volume 4: Earth Science*. ☐ Fill out an Earth Science Record Sheet on SL pg. 32 on ice storms. ☐ Add facts to the Climate Information on SL pg. 29.	☐ Read one of the additional library books. ☐ Go over the vocabulary word and enter it into the Earth Science Glossary on SL pg. 97. ☐ Choose one of the activities for the week to do; fill out the project record sheet on pg. 37.	☐ Do the copywork or dictation assignment and add it to the Earth Science Notes sheet on SL pg. 35. ☐ Check on the rain gauge and add the data to SL pg. 26; Add weather to the Weather Information Sheet on SL pg. 30. ☐ Work on one or all of the multi-week activities.

Science-Oriented Books

Living Book Spine
- 📖 Chapter 6 of *The Sassafras Science Adventures Volume 4: Earth Science*

Optional Encyclopedia Readings
- 🔖 *Basher Science Planet Earth* (No pages scheduled.)
- 🔖 *Usborne Children's Encyclopedia* pg. 15 (Section on Icy Snowflakes)
- 🔖 *Discover Science Weather* pp. 30–31 (Big Freeze)
- 🔖 *Usborne Encyclopedia of Planet Earth* pg. 88 (Freezing)

Additional Living Books
- 📖 *Snow* by Uri Shulevitz
- 📖 *Snow Is Falling (Let's-Read-and-Find... Science, Stage 1)* by Franklyn M. Branley and Holly Keller
- 📖 *Snow (Ready-to-Reads)* by Marion Dane Bauer and John Wallace

Notebooking (SCIDAT Logbook Information)

This week, you can have the students begin to fill out the Climate Sheet for the Patagonia. They can also fill out the first part of their weather record sheets and the logbook sheets for snow and ice storms. Here is the information they could include:

Climate Sheets

Area Map – Have the students color the Patagonia region at the tip of South America. This region is found in both Chile and Argentina.

Climate Information – This week, the students could include the following:
⇨ There is quite a bit of winds in the Patagonia region, which can make it feel colder than it really is. This phenomenon is known as wind chill. Wind chill is when the temperature your body feels is the air temperature plus the wind speed.

Interesting Facts – Answers will vary.

Other Types – This week, the students could include the following:
⇨ The northern portion of the Patagonia region is part of the pampas, or grasslands, of Argentina. The southern part transitions from the pampas into to the taiga. It is much cooler and frosts can occur throughout the year.

Weather Record Sheets

Have the students record the weather from your area or from Patagonia over the week.

Earth Science Record Sheets

Snow

Information Learned
⇨ Snow forms by a process of deposition, which means that water vapor in the high in the atmosphere changes directly into ice without becoming a liquid first. The temperature must be below freezing (32°F) for this to occur.

⇨ *If snow meets any warm air as it falls to the ground, it can be turned into rain, sleet, or freezing rain.*
⇨ *Snowflakes come in many shapes and sizes, but each one is six-sided.*
⇨ *Snowflakes form from as many as two hundred ice crystals that come together in a lattice structure around a tiny piece of dust or dirt.*
⇨ *Snow is white because the crystalline structure reflects all light, making it appear white to our eyes.*
⇨ *A blizzard is a long-lasting snow storm with strong winds. Blizzards can deposit a large amount of snow in a relatively short amount of time.*

Ice storms
Information Learned

⇨ *An ice storm is a winter storm with freezing rain instead of snow.*
⇨ *Ice storms occur when there is a layer of warm air sandwiched by two layers of cold air. The snow falls, melts, and then refreezes as ice as it reaches the surface.*
⇨ *Freezing rain occurs when the air close the surface is below freezing, so that the melted snow refreezes when it hits the surface.*
⇨ *Sleet is raindrops that freeze before they reach the ground. Sleet doesn't typically stick to surfaces, but it can accumulate just like snow.*

Vocabulary

Have the older students look up the following term in the glossary in the Appendix on pp. 109-110 or in a science encyclopedia. Then, have them copy the definition onto a blank index card or into their SCIDAT logbook.

☁ **SNOWFLAKE** – A collection of ice crystals that form a shape with six similar sides.

Scientific Demonstration: Snowflakes

Materials
- ☑ Glass Jar
- ☑ 2 Pipe cleaners
- ☑ Pencil
- ☑ Borax (**NOTE** – *This can be found in the laundry detergent aisle of the grocery store.*)
- ☑ Water

Procedure

1. Add 1 cup of hot water to the jar. Then, add 3 tablespoons of Borax, one tablespoon at a time. Taking care each time to stir until the Borax is dissolved. (**NOTE** – *The jar will get very warm, so do not leave the students unsupervised at any point during this demonstration!!*)
2. Meanwhile, have the students shape a snowflake out of one of the pipe cleaners. This can be as simple or as complex as they wish, but make sure it will fit through the opening of your jar. Then, have them attach one end of the second pipe cleaner to the center of the snowflake and the other end to the center of the pencil.
3. Have the students hang their snowflake in the jar so that it is completely covered by the liquid, but not touching the bottom. Allow the jar to sit undisturbed overnight.
4. The next morning, observe the changes that have occurred in the jar.

Explanation

The students should see that the pipe cleaner snowflake is completely coated with crystals. The crystals should be white, making the pipe cleaner snowflake look like ice or snow. Hot water can hold more Borax crystals than cold water. This is because the hot water molecules are further apart making more room in which the Borax crystals can dissolve. In this way, you were able to super saturate the water with Borax crystals. As the water cooled, the molecules came closer together, causing the Borax crystals to come out of solution and form crystals again. The pipe cleaners give the crystals an easy place to attach and begin to form, which is why they are coated with the crystals.

Take it further
Have the students put their snowflakes back into the jar and let the jars sit undisturbed for an additional two days to see if any more crystals are formed.

Multi-Week Projects and Activities

Multi-week Projects
- **Weather Poster** – Have the students add to their weather poster this week. (*Weather template pictures can be found in the Appendix on pg. 8.*)

Activities For This Week
- **Indoor Snow** – You will need a box of cornstarch and a can of shaving cream. Have the students mix equal parts of cornstarch and shaving cream to create snow. Let them play to their hearts' content!
- **Ice Painting** – You will need Epsom salts, hot water, food coloring, and paper. Mix equal parts of the Epsom salts and hot water together until most of the Epsom salts have dissolved. Add a few drops of food coloring and mix well. Then, have the students use the mixture to paint a snowflake design or ice storm on the paper. As it dries, the ice crystals will form!
- **Microscope Work** – If you are lucky enough to have snow at the time you are doing this unit, have the students look at snow under the microscope. You will need to place the slide in the freezer for thirty minutes so that the snow won't melt as soon as it hits the slide. Head outside and gather a few snowflakes on the slide, put them under the microscope, and view immediately as the light will quickly melt the snow. Have the students complete one of the microscope worksheets found on pp. 97-98 of the Appendix. If you don't have snow or a microscope, you can view snow under an electron microscope at the following website:
  - http://www.designsoak.com/snowflakes-under-microscope/

Memorization

Copywork/Dictation

☞ **Copywork Sentence**
Snowflakes come in many shapes and sizes, but each one is six-sided.

☞ **Dictation Selection**
Snow forms by a process of deposition, which means that water vapor in the high in the atmosphere changes directly into ice without becoming a liquid first. The temperature must be below freezing for this to occur. Snowflakes form from as many as two hundred ice crystals that come together in a lattice structure around a tiny piece of dust or dirt.

Notes

Chapter 7: Out of the Office

Chapter Summary

The chapter opens with the group embarking on their third challenge for the Out of the Office show. This one involves making a fire and this time Ted, one of the competing Q. B. Cubicles office workers, wins. The group warms themselves by the fire and learns how to make a shelter out of evergreen branches before heading to bed for the night. They are awakened by a loud cracking and shaking of the ground, which Hawk Talons informs them is the result of a frost quake. While awake, the challengers observe the beauty of the southern lights before heading back to bed. The next morning, the fourth challenge is issued—find food. The challengers head out and we follow Tracey, who had an idea for finding fish. She thinks that she is followed by the Man with No Eyebrows, but it turns out to be a camera man who films her as she captures food for the group. The group eats and then begins to follow the stream Tracey found to find a village. On the way, Hawk tells them about the different seasons.

Supplies Needed

Demonstration	Projects and Activities
• Small paper cup, Water	• Air dry clay, Brown pipe cleaners, Felt • Clear glass, Crushed ice, Salt

Optional Schedule for Two-Days-A-Week

Day 1	Day 2
☐ Read the section entitled "Frost Quake!" of Chapter 7 in *SSA Volume 4: Earth Science*.	☐ Read the section entitled "Seasonal Shifts" of Chapter 7 in *SSA Volume 4: Earth Science*.
☐ Fill out an Earth Science Record Sheet on SL pg. 33 on frost quakes.	☐ Fill out an Earth Science Record Sheet on SL pg. 34 on the seasons.
☐ Add facts to the Climate Information on SL pg. 29; Add weather to the Weather Information Sheet on SL pg. 30.	☐ Add weather to the Weather Information Sheet on SL pg. 30; Check on the rain gauge and add the data to SL pg. 26.
☐ Do the demo entitled "Expanding Ice"; write information learned on SL pg. 36.	☐ Go over the vocabulary words and enter them into the Earth Science Glossary on SL pg. 97.
☐ Do the copywork or dictation assignment and add it to the Earth Science Notes on SL pg. 36.	☐ Work on one or all of the multi-week activities.

Optional Schedule for Five-Days-A-Week

Day 1	Day 2	Day 3	Day 4	Day 5
☐ Read the section entitled "Frost Quake!" of Chapter 7 in *SSA Volume 4: Earth Science*. ☐ Fill out an Earth Science Record Sheet on SL pg. 33 on frost quakes. ☐ Add weather to the Weather Information Sheet on SL pg. 30.	☐ Read one or all of the assigned pages from the encyclopedia of your choice; write narration on the Earth Science Notes Sheet on SL pg. 36. ☐ Do the demo entitled "Expanding Ice"; write information learned on SL pg. 36.	☐ Read the section entitled "Seasonal Shifts" of Chapter 7 in *SSA Volume 4: Earth Science*. ☐ Fill out an Earth Science Record Sheet on SL pg. 34 on the seasons. ☐ Add facts to the Climate Information on SL pg. 29.	☐ Read one of the additional library books. ☐ Go over the vocabulary words and enter them into the Earth Science Glossary on SL pg. 97. ☐ Choose one of the activities for the week to do; fill out the project record sheet on pg. 38.	☐ Do the copywork or dictation assignment and add it to the Earth Science Notes sheet on SL pg. 36. ☐ Check on the rain gauge and add the data to SL pg. 26; Add weather to the Weather Information Sheet on SL pg. 30. ☐ Work on one or all of the multi-week activities.

Science-Oriented Books

Living Book Spine
- Chapter 7 of *The Sassafras Science Adventures Volume 4: Earth Science*

Optional Encyclopedia Readings
- *Basher Science Planet Earth* pp. 76 *(Seasons)*
- *Discover Science Weather* pp. 12-13 *(Changing Seasons)*
- *Usborne Children's Encyclopedia* pp. 12-13 *(The Seasons)*
- *Usborne Encyclopedia of Planet Earth* pp. 12-13 *(The Seasons)*

Additional Living Books
- *Tree For All Seasons (Avenues)* by Robin Bernard
- *On the Same Day in March: A Tour of the World's Weather* by Marilyn Singer and Frane Lessac
- *Sunshine Makes the Seasons (Let's-Read-and-Find... Science 2)* by Franklyn M. Branley and Michael Rex
- *Watching the Seasons (Welcome Books)* by Edana Eckart
- *The Reasons for Seasons* by Gail Gibbons

Notebooking (SCIDAT Logbook Information)

This week, you can have the students add to the Climate Sheet for Patagonia. They can also fill out the second part of their weather record sheets and the logbook sheets for frost quakes and the seasons. Here is the information they could include:

Climate Sheets
Climate Information – This week, the students could include the following:
⇨ *Spring, summer, and fall are relatively short in Patagonia. Winter is the longest season, but when it is winter there, it is summer in the Northern hemisphere.*

Interesting Facts – Answers will vary. Students can add the following facts about the Aurora Australias:
⇨ *The Aurora Australias is known as the Southern Lights. These lights are an astronomical phenomenon that look like shafts or curtains of light in the night sky during the winter months. Similar lights are also found in the northern hemisphere, where they are known as the Aurora Borealis.*

Other Types – This week, the students could include the following:
⇨ *The northern portion of the Patagonia region is part of the pampas, or grasslands, of Argentina. The southern part transitions from the pampas in to the taiga. It is much cooler and frosts can occur throughout the year.*

Weather Record Sheets
Have the students record the weather from your area or from Patagonia over the week.

Earth Science Record Sheets
Frost Quakes
Information Learned
⇨ *Frost quakes occur in the middle of the night when the temperatures are at their coldest.*
⇨ *As the temperature cools, the water vapor in the air close to the ground condenses,*

freezes, and forms ice crystals. It appears on the ground and other surfaces when the temperature falls below freezing. If the temperature does not go below freezing, dew forms instead.
- ⇨ They are not caused by a shift of the earth's plates; rather, they are caused by a sudden rapid freezing of the ground or by glacial movement.
- ⇨ For a frost quake to occur, water needs to saturate the soil and fill gaps in the rocks or in the glacier. Usually this happens when the temperature goes above freezing and some of the surface snow or ice melts. Then, the temperature needs to plummet quickly below freezing again, causing that water to rapidly freeze and expand. The pressure cause the soil, rock, or glacier to crack in an explosive manner, which creates a loud sound and can even shake the surrounding ground.
- ⇨ Frost quakes are also known as cryoseisms.

Seasons
Information Learned
- ⇨ We have four seasons – spring, summer, autumn (or fall), and winter.
- ⇨ A season is a collection of days with a typical weather pattern.
- ⇨ They are a result of the changing position of the earth as it rotates around the sun and the tilt of the earth on its axis.
- ⇨ The earth's orbit around the sun is slightly elliptical, so during certain times of the year, the earth is closer to the sun than at other times, which make the weather warmer than during other times of the year.
- ⇨ Since the earth is slightly tilted on its axis, sometimes one hemisphere it is tilted towards the sun, making the weather warmer, while the other hemisphere is tilted away from the sun, making the weather colder. This is why when it is summer in the northern hemisphere, it is winter in the southern hemisphere, and vice versa.

Vocabulary
Have the older students look up the following terms in the glossary in the Appendix on pp. 109-110 or in a science encyclopedia. Then, have them copy each definition onto a blank index card or into their SCIDAT logbook.
- ✏ EQUATOR – An imaginary line that divides the earth into the Northern and Southern Hemisphere.
- ✏ SEASON – A collection of days with a typical weather pattern.

Scientific Demonstration: Expanding Ice
Materials
- ☑ Small paper cup
- ☑ Water

Procedure
1. Have the students fill small paper cup to the top and put the cup in the freezer where it won't be disturbed.
2. Have the students check the cup after two hours to observe what has happened.
   NOTE – *If you have older students, have them measure and record the amount of water beforehand. Then, let the ice melt and have the students measure and record the amount once more to see if there is a difference.*

Explanation
The students should see that the water has turned into ice and that the ice has moved up and over the top of the cup. This is because water expands as it freezes. This principle can lead to frost quakes and rock weathering.

Take it Further
Have the students repeat the demonstration, only this time have the students put a lid on their cup to see if the results change.

Multi-Week Projects and Activities
Multi-week Projects
✂ **WEATHER POSTER** – Have the students add to their weather poster this week. (*Weather template pictures can be found in the Appendix on pg. 8.*)

Activities for This Week
✂ **SEASONS** – Have the students make a tree from a season of their choice. You will need a lump of air dry clay, several brown pipe cleaners, and felt sheets of either green, red, orange, or yellow. Have the students begin by shaping the pipe cleaners for the trunk and branches. Once they are finished, have them secure the tree into a lump of clay as a base. Then, have them pick a season and cut out felt leaves in the right color. In other words, if they chose to make a spring tree, they would use light green felt for their leaves. Have the students glue these onto the tree. Finally, have them decorate their tree with things like flowers, bird's nests, and fruit or nuts. For example, for the spring tree, the students could add small pieces of crushed tissue paper for flowers.

✂ **FROSTY MUG** – Have the students make some frost! You will need a clear glass, crushed ice, and salt. Have the students fill the glass halfway with crushed ice. Then, have them add four to five teaspoons of salt and mix well. As the salt melts, the student swill see frost forming on the outside of the glass. (*This is because the salt water is actually below the freezing point for water, so when the water vapor comes in contact with the glass, it condenses and freezes on the surface.*)

✂ **NORTHERN/SOUTHERN LIGHTS** – Have the students watch the following video to learn more about this spectacular phenomenon:
  🖱 https://www.youtube.com/watch?v=eJV_wlCm6ms

Memorization
Copywork/Dictation
☞ **COPYWORK SENTENCE**
  We have four seasons – spring, summer, autumn (or fall), and winter.

☞ **DICTATION SELECTION**
  The Four Seasons
  The short days of winter are filled with cold and snow
  In spring, it warms as the flowers begin to grow
  The long days of summer are hot and green
  In fall, the changing leaves create a beautiful scene

Notes

CHAPTER 8: THE GOBI DESERT

CHAPTER SUMMARY

The chapter opens with Blaine and Tracey zipping to Mongolia, where they land in the middle of nowhere. Thankfully, a man shortly rides up on horseback, who turns out to be their local expert, Ganzorig Buri. He shares a bit about the desert and the day/night cycle before inviting the twins to join him on his journey back to his village. On their way, they run into his dear friend, Solongo. She tells the group about how a band of Avargatom warriors are terrorizing the village. The group races homeward with a sandstorm on their heels. Ganzorig shares about sandstorms as they rush to find shelter. Once the storm passes over, the travelers continue their journey.

SUPPLIES NEEDED

Demonstration	Projects and Activities
• A globe (or large ball), A desk lamp • A Post-it tab (or another type of removable marker)	• Flour, Vegetable oil • Microscope slides, Sand

OPTIONAL SCHEDULE FOR TWO-DAYS-A-WEEK

Day 1	Day 2
☐ Read the section entitled "Is is day or night?" of Chapter 8 in *SSA Volume 4: Earth Science*.	☐ Read the section entitled "Swirling Sandstorms" of Chapter 8 in *SSA Volume 4: Earth Science*.
☐ Fill out an Earth Science Record Sheet on SL pg. 41 on the day/night cycle.	☐ Fill out an Earth Science Record Sheet on SL pg. 42 on sandstorms.
☐ Add weather to the Weather Information Sheet on SL pg. 40.	☐ Add facts to the Climate Information on SL pg. 39; Add weather to the Weather Information Sheet on SL pg. 40.
☐ Go over the vocabulary word and enter it into the Earth Science Glossary on SL pg. 98.	☐ Do the demo entitled "Day or Night"; write information learned on SL pg. 45.
☐ Do the copywork or dictation assignment and add it to the Earth Science Notes on SL pg. 45.	☐ Work on one or all of the multi-week activities.

OPTIONAL SCHEDULE FOR FIVE-DAYS-A-WEEK

Day 1	Day 2	Day 3	Day 4	Day 5
☐ Read the section entitled "Is it day or night?" of Chapter 8 in *SSA Volume 4: Earth Science*.	☐ Read the section entitled "Swirling Sandstorms" of Chapter 8 in *SSA Volume 4: Earth Science*.	☐ Read one or all of the assigned pages from the encyclopedia of your choice; write narration on the Earth Science Notes Sheet on SL pg. 45.	☐ Read one of the additional library books.	☐ Do the copywork or dictation assignment and add it to the Earth Science Notes sheet on SL pg. 45.
☐ Fill out an Earth Science Record Sheet on SL pg. 41 on the day/night cycle.	☐ Fill out an Earth Science Record Sheet on SL pg. 42 on sandstorms.	☐ Do the demo entitled "Day or Night"; write information learned on SL pg. 45.	☐ Go over the vocabulary word and enter it into the Earth Science Glossary on SL pg. 98.	☐ Add weather to the Weather Information Sheet on SL pg. 40.
☐ Add weather to the Weather Information Sheet on SL pg. 40.	☐ Add facts to the Climate Information on SL pg. 39.		☐ Choose one of the activities for the week to do; fill out the project record sheet on pg. 47.	☐ Work on one or all of the multi-week activities.

The Sassafras Guide to Earth Science ~ Chapter 8

Science-Oriented Books

Living Book Spine
- Chapter 8 of *The Sassafras Science Adventures Volume 4: Earth Science*

Optional Encyclopedia Readings
- *Basher Science Planet Earth* (No pages scheduled.)
- *Usborne Children's Encyclopedia* pp. 10-11 (Day and Night)
- *Discover Science Weather* (No pages scheduled.)
- *Usborne Encyclopedia of Planet Earth* pp. 14-15 (Day and Night)

Additional Living Books
- *Day and Night (First Step Nonfiction: Discovering Nature's Cycles)* by Robin Nelson
- *Day and Night (Patterns in Nature)* by Margaret Hall and Jo Miller
- *What Makes Day and Night (Let's-Read-and-Find... Science 2)* by Franklyn M. Branley and Arthur Dorros
- *Extreme Weather: Surviving Tornadoes, Sandstorms, Hailstorms, Blizzards, Hurricanes, and More!* by Thomas M. Kostigen

Notebooking (SCIDAT Logbook Information)

This week, you can have the students begin to fill out the Climate Sheet for the Mongolian Desert. They can also fill out the first part of their weather record sheets and the logbook sheets for the day/night cycle and sandstorms. Here is the information they could include:

Climate Sheets

Area Map – Have the students color the region where the Mongolian Desert is found. (*See map for answers.*)

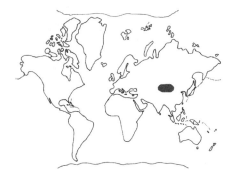

Climate Information – This week, the students could include the following:
- ⇨ *The Mongolian Desert (a.k.a., the Gobi Desert) climate is known as a continental climate because it is not influenced by a neighboring sea – it is too far inland.*
- ⇨ *The climate in the Gobi desert is very extreme – it can reach over 100°F in the summer and -40°F in the winter.*

Interesting Facts – Answers will vary. Students can add the following fact:
- ⇨ *The Mongolian Desert (a.k.a., the Gobi Desert) is known as the land of the blue sky because clouds rarely are seen.*
- ⇨ *A mirage is when one sees a pool of water that is not there. It happens because of the temperature difference between the ground and air. The ground is very hot and the air is cool, which causes light to be refracted in such a way that it looks like a pool of water.*

Other Types – This week, the students could include the following:
- ⇨ *Hot deserts are found around the center of the Earth. The climate is typically very hot during the day and cool at night year-round. There is very little rain in the hot desert.*
- ⇨ *Cold deserts around found north and south of the center of the Earth. The climate is very hot during the summer and very cold during the winter. Cold deserts get some of the minuscule rainfall in the form of snow.*

Weather Record Sheets
Have the students record the weather from your area or from the Gobi Desert over the week.

Earth Science Record Sheets
Day / Night
Information Learned
⇨ As the Earth turns, or rotates, day changes to night on one side of the globe.
⇨ On the other side, night changes to day with the rotation of our planet. This is because as the Earth turns, different parts of the surface face the Sun.
⇨ In the morning, our side of the Earth is turning to face the sun. This is when we see the sunrise, which ushers in a new day.
⇨ In the evening, our side of the Earth is rotating away from the Sun. This is when we see the sunset, which marks the beginning of a new night.
⇨ The day-night cycle takes a full twenty-four hours to complete.

NAME: Day and Night
INFORMATION LEARNED: The day/night cycle takes a full twenty-four hours to complete.

Sandstorm
Information Learned
⇨ Sandstorms are strong, violent winds that stir up loose sand and sediment, carrying it to another location.
⇨ Sandstorms are caused by a front that moves through an arid desert region. These storms can occur around the globe and they can be small or extremely large. A sandstorm can spread over hundreds of miles.
⇨ Sandstorms can be devastating and dangerous, but they can also be beneficial. For instance, sandstorms in the Sahara carry near twenty tons of particles all the way to the Amazon River basin, which provides the forest with the essential nutrients and minerals the area needs.
⇨ Sandstorms often arrive with little to no warning and usually blow through an area quickly, in five minutes or so.
⇨ The dust carried by a sandstorm can cause blinding conditions and after a storm has blown through, the devastation can look similar to a mild tornado. Sandstorms can also pick up spores, bacteria, or pollutants and carry them hundreds of miles away.

NAME: Sandstorm
INFORMATION LEARNED: Sandstorms are strong violent winds that stir up loose sand and sediment.

Vocabulary
Have the older students look up the following term in the glossary in the Appendix on pp. 109-110 or in a science encyclopedia. Then, have them copy the definition onto a blank index card or into their SCIDAT logbook.

✏ SANDSTORM – A strong, violent winds that stir up loose sand and sediment, carrying it to another location.

Scientific Demonstration: Day or Night
Materials
☑ A globe (or large ball)
☑ A desk lamp
☑ A Post-it tab (or another type of removable marker)

Procedure
1. Have the students mark where they live on the globe with the Post-it tab.
2. Then, have them place the globe on a desk and shine the desk lamp on the portion of the globe where they live.
3. Have the students slowly spin (or rotate) the globe to see what happens.

Explanation

The students should see that as they rotate the globe, the place where they live rotates out of the light. As they continue to spin the globe, the place where they live will rotate back into the light – creating a full day/night cycle..

Take it Further

Have the students draw their own houses during the day and at night. Then, have them write the different activities they do during the day and at night on the bottom of the page.

Multi-Week Projects and Activities

Multi-week Projects

- **Weather Poster** – Have the students add to their weather poster this week. (*Weather template pictures can be found in the Appendix on pg. 8.*)

Activities For This Week

- **Day & Night Song** – Have the students watch and sing along with this cute song about what makes it day and night:
  - https://www.youtube.com/watch?v=ZoG1pF_r5zU

  Or watch the following video on the day/night cycle:
  - http://youtu.be/hWkKSkI3gkU

- **Desert Sand** – Mix up a batch of desert sand for the students to play with. You will need flour and vegetable oil. Measure out 4 cups of flour and mix in ½ cup of oil—the mixture will be dry and crumbly, sort of like sand. Let the students play and enjoy!

- **Microscope Work** – Have the students look at what is found in the sand under the microscope. They can do this by sprinkling a bit of sand on a slide before bringing the slide inside to view under the microscope. Have the students complete one of the microscope worksheets found on pp. 97-98 of the Appendix. If you do not have a microscope, you can view several images of sand under a microscope here:
  - http://www.microlabgallery.com/SandOregonFile.aspx

Memorization

Copywork/Dictation

☞ **Copywork Sentence**

The day-night cycle takes a full twenty-four hours to complete.

☞ **Dictation Selection**

As the Earth turns, or rotates, day changes to night on one side of the globe. On the other side, night changes to day with the rotation of our planet. This is because as the Earth turns, different parts of the surface face the Sun. The day-night cycle takes a full twenty-four hours to complete.

Notes

Chapter 9: Avargatom Challenges

Chapter Summary

The chapter opens with Blaine, Tracey, Solongo, and Ganzorig refreshing themselves at a local well. The group then makes their way to the Buri village, where they find the leader of the Avagartom issuing forth challenges that the tribesmen are reluctant to accept. Ganzorig shares with the twins about droughts as he watches his brother, Dariin, compete to preserve the village's food supply. Unfortunately, Dariin looses, but the Avagartom issue a second challenge—one in which the Buri will either get some of their food back or lose their tools. Ganzorig's other brother, Khulan, accepts the challenge, but loses. A third and final challenge is issued and this time Ganzorig steps up. He ups the ante by demanding that the contest be for possession of the Avagartom oasis and the Buri village. The twins learn about oases from Solongo as they watch their local expert win the competition! The chapter ends with us finding out the Man with No Eyebrows is heading out to meet the twins at their next location.

Supplies Needed

Demonstration	Projects and Activities
• Dark construction paper, Water, Salt, Eyedropper	• Materials will vary

Optional Schedule for Two-Days-A-Week

Day 1	Day 2
☐ Read the section entitled "Desperate Drought" of Chapter 9 in *SSA Volume 4: Earth Science*.	☐ Read the section entitled "Oasis in the Trials" of Chapter 9 in *SSA Volume 4: Earth Science*.
☐ Fill out an Earth Science Record Sheet on SL pg. 43 on droughts.	☐ Fill out an Earth Science Record Sheet on SL pg. 44 on oases.
☐ Add facts to the Climate Information on SL pg. 39; Add weather to the Weather Information Sheet on SL pg. 40.	☐ Add weather to the Weather Information Sheet on SL pg. 40; Go over the vocabulary word and enter it into the Earth Science Glossary on SL pg. 98.
☐ Do the copywork or dictation assignment and add it to the Earth Science Notes on SL pg. 46.	☐ Do the demo entitled "Drought Plate"; write information learned on SL pg. 46.
	☐ Work on one or all of the multi-week activities.

Optional Schedule for Five-Days-A-Week

Day 1	Day 2	Day 3	Day 4	Day 5
☐ Read the section entitled "Desperate Drought" of Chapter 9 in *SSA Volume 4: Earth Science*.	☐ Read the section entitled "Oasis in the Trials" of Chapter 9 in *SSA Volume 4: Earth Science*.	☐ Read one or all of the assigned pages from the encyclopedia of your choice; write narration on the Earth Science Notes Sheet on SL pg. 46.	☐ Read one of the additional library books.	☐ Do the copywork or dictation assignment and add it to the Earth Science Notes sheet on SL pg. 46.
☐ Fill out an Earth Science Record Sheet on SL pg. 43 on droughts.	☐ Fill out an Earth Science Record Sheet on SL pg. 44 on oases.	☐ Do the demo entitled "Drought Plate"; write information learned on SL pg. 46.	☐ Go over the vocabulary word and enter it into the Earth Science Glossary on SL pg. 98.	☐ Add weather to the Weather Information Sheet on SL pg. 40.
☐ Add facts to the Climate Information on SL pg. 39.	☐ Add weather to the Weather Information Sheet on SL pg. 40.		☐ Choose one of the activities for the week to do; fill out the project record sheet on pg. 48.	☐ Work on one or all of the multi-week activities.

Science-Oriented Books

Living Book Spine
- Chapter 9 of *The Sassafras Science Adventures Volume 4: Earth Science*

Optional Encyclopedia Readings
- *Basher Science Planet Earth* pg. 90 (Drought), pp. 112-113 (Desert)
- *Discover Science Weather* pp. 28-29 (Wet and Dry)
- *Usborne Children's Encyclopedia* pp. 28-29 (In the desert)
- *Usborne Encyclopedia of Planet Earth* pg. 87 (Droughts), pp. 64-65 (Tropical Deserts)

Additional Living Books
- *Droughts (Weather Update)* by Nathan Olson
- *Droughts (Blastoff Readers Level 4)* by Anne Wendorff
- *A Desert Habitat (Introducing Habitats)* by Kelley Macaulay and Bobbie Kalman
- *About Habitats: Deserts* by Cathryn P. Sill
- *Life in the Desert (Pebble Plus: Habitats Around the World)* by Alison Auch

Notebooking (SCIDAT Logbook Information)

This week, you can have the students add to the Climate Sheet for the Mongolian Desert. They can also fill out the second part of their weather record sheets and the logbook sheets for droughts and oases. Here is the information they could include:

Climate Sheets
There is no information for the students to add to the climate sheets for this week.

Weather Record Sheets
Have the students record the weather from your area or from the Gobi Desert over the week.

Earth Science Record Sheets

Drought
Information Learned
- A drought is an abnormally long period of time (weeks, months, or even years) with little to no precipitation of any kind.
- During a drought, the conditions are typically hot, dry, and dusty.
- The word drought can be used to refer to many conditions, so scientists got together and created four areas or types of drought. These four types are:
    1. Meteorological – This type of drought is caused by the lack of precipitation accompanied by winds and high temperatures. Water sources dry up, ground water is reduced, and the soil can become as hard as a rock and crack.
    2. Agricultural – This type of drought happens when the soil dries out so much that crops and animals are affected.
    3. Hydrological – This type of drought is a reduction of surface or ground water due to overuse.
    4. Socioeconomic – This type of drought results in the lack of services, such as drinking water and

electricity, due to meteorological or hydrological conditions.

Oasis
Information Learned

- ➭ An oasis is a place in the desert where a pool of water can be found. The pool is formed when water rises from underneath the ground to form a spring. This can occur naturally or can be the result of a man-made well.
- ➭ Man-made oases are typically found sprinkled across the old trade routes, part of the Silk Road that went through the Gobi desert.
- ➭ Oases can be very small, with only a cluster of small plants, or quite large.
- ➭ In the Gobi desert there, are several underground rivers that feed springs that create oases.
- ➭ Thanks to the water, plants can grow in that area of the desert. Many of the desert animals gather in these regions to drink, eat, and find shade.

Vocabulary

Have the older students look up the following term in the glossary in the Appendix on pp. 109-110 or in a science encyclopedia. Then, have them copy the definition onto a blank index card or into their SCIDAT logbook.

- ⚅ DROUGHT – A long period without rain.

Scientific Demonstration: Drought Crust
Materials
- ☑ Dark construction paper (black or brown will work best)
- ☑ Water
- ☑ Salt
- ☑ Eyedropper

Procedure
1. Have the students mix ¼ cup of warm water with 2 teaspoons of salt. Stir until most of the salt has dissolved.
2. Next, find a spot inside that is in direct sunlight and will remain in the light for at least two hours. Have the students set the paper in the spot.
3. Then, have the students use the eyedropper to drip the saltwater in spots all over the paper.
4. Check the plate every fifteen minutes over the next two hours and observe the changes that occur.

Explanation

The students should see that the drops on the paper have dried and that a crust of salt has been left behind on top of the paper. The sun has dried out the surface, causing the salt to crystallize and create a crust, much like what occurs on the surface of the soil during a drought.

Take it Further

Have the students repeat the demonstration with different types of soil to see the effect it has on the results. It will take longer for the soil to dry completely, so be patient.

Multi-Week Projects and Activities
Multi-week Projects
- ✂ WEATHER POSTER – Have the students add to their weather poster this week. (*Weather template pictures can be found in the Appendix on pg. 8.*)

Activities For This Week
- ✂ WATER CONSERVATION – Have the students learn more about how they can help conserve water

with the following video:
- 🖱 https://www.youtube.com/watch?v=0Am9JPfuNsw
- ✂ OASIS – Have the students make their own replica of an oasis. You can have them create a salt dough map, a habitat diorama, or simply paint a poster of an oasis.

Memorization

Copywork/Dictation

☞ **Copywork Sentence**

A drought is a long period without rain.

☞ **Dictation Selection**

An oasis is a place in the desert where a pool of water can be found. The pool is formed when water rises from underneath the ground to form a spring. This can occur naturally or can be the result of a man-made well. Man-made oases are typically found sprinkled across the old trade routes, part of the Silk Road that went through the Gobi desert.

Notes

Chapter 10: Wolves in Pakistan

Chapter Summary

The chapter opens with the Man with No Eyebrows hiding out in the sheep pen so that he can surprise Blaine and Tracey once they arrive in Pakistan. He is found out just before the Sassafras twins arrive. The twins spend the night among the sheep and in the morning they are awakened up by a group of children. Tariq, Javeria, and Aazmi are all apprentices of the Shepherd, who turns out to be the twins' local expert. They join his journey to find the lost sheep and learn about the atmosphere on the way. After that, we learn that the Man with No Eyebrows has missed the twins and returned to his lab. We rejoin the twins and their companions as they make their way up the mountain. As they follow the tracks, the children learn about clouds and the Shepherd discovers that the infamous Raider is the most likely suspect to have taken his sheep. The chapter ends with Blaine losing his footing as he is distracted by a stack of pancake-shaped clouds!

Supplies Needed

Demonstration	Projects and Activities
• Clear plastic cup, Soda bottle, Blue food coloring • Water, Marker	• White paint, Cotton balls, Blue construction paper

Optional Schedule for Two-Days-A-Week

Day 1	Day 2
☐ Read the section entitled "Ascending through the…" of Chapter 10 in *SSA Volume 4: Earth Science*. ☐ Fill out an Earth Science Record Sheet on SL pg. 51 on the atmosphere. ☐ Add weather to the Weather Information Sheet on SL pg. 50. ☐ Do the demo entitled "Barometer"; write information learned on SL pg. 57. ☐ Do the copywork or dictation assignment and add it to the Earth Science Notes on SL pg. 57.	☐ Read the section entitled "Capturing Clouds" of Chapter 10 in *SSA Volume 4: Earth Science*. ☐ Fill out an Earth Science Record Sheet on SL pg. 52 on clouds. ☐ Add facts to the Climate Information on SL pg. 49; Add weather to the Weather Information Sheet on SL pg. 50. ☐ Go over the vocabulary words and enter them into the Earth Science Glossary on SL pg. 98. ☐ Work on one or all of the multi-week activities.

Optional Schedule for Five-Days-A-Week

Day 1	Day 2	Day 3	Day 4	Day 5
☐ Read the section entitled "Ascending through the…" of Chapter 10 in *SSA Volume 4: Earth Science*. ☐ Fill out an Earth Science Record Sheet on SL pg. 51 on the atmosphere. ☐ Add weather to the Weather Information Sheet on SL pg. 50.	☐ Read one or all of the assigned pages from the encyclopedia of your choice; write narration on the Earth Science Notes Sheet on SL pg. 57. ☐ Do the demo entitled "Barometer"; write information learned on SL pg. 57.	☐ Read the section entitled "Capturing Clouds" of Chapter 10 in *SSA Volume 4: Earth Science*. ☐ Fill out an Earth Science Record Sheet on SL pg. 52 on clouds. ☐ Add facts to the Climate Information on SL pg. 49.	☐ Read one of the additional library books. ☐ Go over the vocabulary words and enter them into the Earth Science Glossary on SL pg. 98. ☐ Choose one of the activities for the week to do; fill out the project record sheet on pg. 59.	☐ Do the copywork or dictation assignment and add it to the Earth Science Notes sheet on SL pg. 57. ☐ Add weather to the Weather Information Sheet on SL pg. 50. ☐ Work on one or all of the multi-week activities.

Science-Oriented Books

Living Book Spine
- Chapter 10 of *The Sassafras Science Adventures Volume 4: Earth Science*

Optional Encyclopedia Readings
- *Basher Science Planet Earth* pg. 74 (Atmosphere)
- *Usborne Children's Encyclopedia* pg. 9 (Section on atmosphere)
- *Discover Science Weather* pp. 10-11 (Blanket of Air)
- *Usborne Encyclopedia of Planet Earth* pp. 48-49 T(he Earth's Atmosphere)

Additional Living Books
- *Grandpa, What is Air? (Popular Science for Children)* by Daniel Levy, Efraim Perlmutter and Yona
- *Atmosphere: Air Pollution and Its Effects (Our Fragile Planet)* by Dana Desonie
- *Little Cloud* by Eric Carle

Notebooking (SCIDAT Logbook Information)

This week, you can have the students begin to fill out the Climate Sheet for the Pakistani Mountains. They can also fill out the first part of their weather record sheets and the logbook sheets for atmosphere and clouds. Here is the information they could include:

Climate Sheets
Area Map – Have the students color the region where the mountains are found in Pakistan. (*See map for answers.*)

Interesting Facts – Answers will vary. Students can add the following fact:
⇨ Gravity is the force that pulls everything towards the center of the Earth. It is also the force that holds the atmosphere in place.
⇨ As you get higher in altitude, the effect of gravity is less. We don't really feel that effect, but the atmosphere does. So, the air gets thinner and thinner as you go up, which means that there is less air to breathe.

Weather Record Sheets
Have the students record the weather from your area or from Pakistan over the week.

Earth Science Record Sheets
Atmosphere
Information Learned
⇨ The atmosphere is a blanket of gas that surrounds and protects the planet. It contains the air we breathe. It protects us from being hit by space rocks. The atmosphere traps heat from the sun, which helps to keep the Earth warm.
⇨ The sky looks blue because of the way that sunlight filters through our atmosphere.
⇨ The atmosphere contains five layers:
  1. Troposphere – The layer closest to the Earth. This is the layer of atmosphere where all the weather occurs.
  2. Stratosphere – The next layer of atmosphere is where planes typically fly because the air is very still. The ozone layer is found here. This layer consists

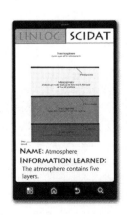

of ozone gas that helps to absorb the harmful ultraviolet rays from the Sun. It is susceptible to damage from certain chemicals that are used in spray cans and refrigerators.

3. *Mesosphere – The next layer of atmosphere is very cool because there are no clouds or ozone to absorb energy from the sun. Many of the rocks from space burn up while passing through this layer.*
4. *Thermosphere – The next layer of atmosphere is very hot due to the presence of atomic oxygen that absorbs energy from the sun. The aurora borealis and australias occur in this layer.*
5. *Exosphere – The farthest layer of atmosphere from the Earth.*

Clouds
Information Learned

⇨ *Clouds are made up of tiny drops of water or ice and dust. They form when warm air holding water vapor cools down.*

⇨ *The way clouds look depends upon how much water is in them and how fast they form. If the clouds form slowly, they typically spread out in sheets. If the clouds form quickly, they puff up into heaps.*

⇨ *Clouds appear white because the water droplets or ice crystals they contain each cause light to be scattered into its different colored components. The net effect combines to be seen as white light.*

⇨ *If clouds get thick enough or full of enough water, not all of the light makes it through, giving them a dark, shadowed appearance.*

⇨ *The darker the clouds are, the larger the droplets of water they contain.*

Vocabulary

Have the older students look up the following terms in the glossary in the Appendix on pp. 109-110 or in a science encyclopedia. Then, have them copy each definition onto a blank index card or into their SCIDAT logbook.

- **ATMOSPHERE** – A layer of gas that surrounds the Earth.
- **VAPOR** – Tiny droplets of water in the air.

Scientific Demonstration: Barometer
Materials
☑ Clear plastic cup
☑ Soda bottle
☑ Blue food coloring
☑ Water
☑ Marker

Trouble shooting Tips – *This demonstration works the best when you set it up on a rainy day, since the air pressure is lower. Also, the bottle needs to be the small enough so that it fits into the cup snugly, but large enough so that it does not touch the bottom.*

Procedure
1. Have the students fill the cup up about a third of the way with water and have them add a few drops of food coloring to the water.
2. Have the students flip the bottle upside down and place it in the water so that the mouth of the bottle is below the waterline, but not touching the bottom of the cup.
3. Use the marker to mark the water level with an inch long thick line on the side of the cup. Then, create a gauge by marking a set of lines, in a quarter of an inch intervals, two inches above and below the line.
4. Have the students set the cup in a spot where it won't be disturbed. Over the next week, have the

students check their barometer daily and observe the position of the water level. (*Older students should record this level on their experiment sheet. They can also choose to record the weather as well.*)

EXPLANATION

The purpose of this demonstration is to allow the students to see and record the pressure changes found in their area. If the students also observe the weather, they should see that the higher the pressure, the clearer the weather.

MULTI-WEEK PROJECTS AND ACTIVITIES

MULTI-WEEK PROJECTS

- **WEATHER POSTER** – Have the students add to their weather poster this week. (*Weather template pictures can be found in the Appendix on pg. 8.*)

ACTIVITIES FOR THIS WEEK

- **ATMOSPHERE VIDEO** – Have the students watch the following video on the layers of the atmosphere:
  - https://www.youtube.com/watch?v=5sg9sCOXFIk
- **ATMOSPHERE POSTER** – Have the students make an atmosphere collage that depicts the different object one can find in the various layers of the atmosphere. Use the information found in chapter 10 or in the suggested encyclopedias for ideas of what to draw. Have the older students write a sentence about what each of the layers are like.
- **CLOUD PAINTINGS** – You will need white paint, cotton balls, and a sheet of blue construction paper. Have the students use the cotton balls to dip in the white paint and then create clouds on their construction paper.

MEMORIZATION

COPYWORK/DICTATION

- **COPYWORK SENTENCE**

  Clouds are made up of tiny drops of water or ice and dust.

- **DICTATION SELECTION**

  The atmosphere is a blanket of gas that surrounds and protects the planet. It contains the air we breathe. It protects us from being hit by space rocks. The atmosphere also traps heat from the sun, which helps to keep the Earth warm. The atmosphere contains five layers: the troposphere, the stratosphere, the mesosphere, the thermosphere, and the exosphere.

NOTES

Chapter 11: The Lost One is Found

Chapter Summary

The chapter opens with Blaine falling, but he is quickly caught by the Shepherd's hook. The group continues its descent down into the town where they believe the Raider has taken the lost sheep. As they go down, their expert shares about cirrus and alto clouds. Once in town, they go to the home of one of the Shepherd's former apprentices, Naveed. They spend the night there and go in search of the Raider at the market in the morning. The Shepherd finds and confronts the Raider, who quickly takes off when he realizes that he has been cornered. The group follows the Raider into a stadium where a polo match is in progress. The Magistrate, the town's governor, halts the game to get to the bottom of the chaos. We find out that Naveed is the one who took the sheep for the Raider and the Shepherd agrees to take the punishment for him. To take their mind off of what is happening, Javeria shares what the Shepherd taught her about stratus and cumulus clouds. The chapter ends with Naveed begging for forgiveness and the Shepherd giving it.

Supplies Needed

Demonstration	Projects and Activities
• Hot water, Glass jar with lid • Crushed ice, Match	• Cotton balls, Blue construction paper

Optional Schedule for Two-Days-A-Week

Day 1	Day 2
☐ Read the section entitled "Out of the Alto Clouds" of Chapter 11 in *SSA Volume 4: Earth Science*.	☐ Read the section entitled "Into the Stratus Market" of Chapter 11 in *SSA Volume 4: Earth Science*.
☐ Fill out an Earth Science Record Sheet on SL pp. 53-54 on cirrus and alto clouds.	☐ Fill out an Earth Science Record Sheet on SL pp. 55-56 on stratus and cumulus clouds.
☐ Add facts to the Climate Information on SL pg. 49; Add weather to the Weather Information Sheet on SL pg. 50.	☐ Add weather to the Weather Information Sheet on SL pg. 50; Go over the vocabulary word and enter it into the Earth Science Glossary on SL pg. 99.
☐ Do the copywork or dictation assignment and add it to the Earth Science Notes on SL pg. 58.	☐ Do the demo entitled "Clouds in a Bottle"; write information learned on SL pg. 58.
	☐ Work on one or all of the multi-week activities.

Optional Schedule for Five-Days-A-Week

Day 1	Day 2	Day 3	Day 4	Day 5
☐ Read the section entitled "Out of the Alto Clouds" of Chapter 11 in *SSA Volume 4: Earth Science*.	☐ Read the section entitled "Into the Stratus Market" of Chapter 11 in *SSA Volume 4: Earth Science*.	☐ Read one or all of the assigned pages from the encyclopedia of your choice; write narration on the Earth Science Notes Sheet on SL pg. 58.	☐ Read one of the additional library books.	☐ Do the copywork or dictation assignment and add it to the Earth Science Notes sheet on SL pg. 58.
☐ Fill out an Earth Science Record Sheet on SL pp. 53-54 on cirrus and alto clouds.	☐ Fill out an Earth Science Record Sheet on SL pp. 55-56 on stratus and cumulus clouds.	☐ Do the demo entitled "Clouds in a Bottle"; write information learned on SL pg. 58.	☐ Go over the vocabulary word and enter it into the Earth Science Glossary on SL pp. 99.	☐ Add weather to the Weather Information Sheet on SL pg. 50.
☐ Add facts to the Climate Information on SL pg. 49.	☐ Add weather to the Weather Information Sheet on SL pg. 50.		☐ Choose one of the activities for the week to do; fill out the project record sheet on pg. 60.	☐ Work on one or all of the multi-week activities.

Science-Oriented Books

Living Book Spine
- 📖 Chapter 11 of *The Sassafras Science Adventures Volume 4: Earth Science*

Optional Encyclopedia Readings
- *Basher Science Planet Earth* pp. 84-85 (Clouds)
- *Discover Science Weather* pp. 22-23 (Mist and clouds)
- *Usborne Children's Encyclopedia* (No pages scheduled.)
- *Usborne Encyclopedia of Planet Earth* pg. 81 (Section on Clouds)

Additional Living Books
- 📖 *Clouds (Let's-Read-and-Find... Science 1)* by Anne Rockwell and Frane Lessac
- 📖 *The Cloud Book* by Tomie dePaola
- 📖 *Clouds (Ready-to-Reads)* by Marion Dane Bauer and John Wallace

Notebooking (SCIDAT Logbook Information)

This week, you can have the students add to the Climate Sheet for the Pakistani Mountains. They can also fill out the second part of their weather record sheets and the logbook sheets for cirrus, alto, stratus, and cumulus clouds. Here is the information they could include:

Climate Sheets
Climate Information – This week, the students could include the following:
- ⇨ *At high altitudes it is typically much cooler because there is less air to trap the heat.*
- ⇨ *Winds are also stronger at higher elevations, and there is a stronger possibility of heavy rain or snow storms.*
- ⇨ *Over 1500 m (4900 ft) humans start to feel the effects of the thinner air at high altitudes. It can cause our heart rates to rise and breathing to increase.*

Weather Record Sheets
Have the students record the weather from your area or from Pakistan over the week.

Earth Science Record Sheets
Cirrus Clouds
Information Learned
- ⇨ *Cirrus clouds are high and wispy.*
- ⇨ *These clouds are found above 18,000 ft and are composed of ice that high winds have blown into long, thin streams.*
- ⇨ *When you see cirrus clouds, it tells you that the weather will be changing in the next twenty-four hours.*
- ⇨ *There are three main types of cirrus clouds:*
    1. *Cirrus – Typically white, wispy and streamer-like clouds; they predict fair and pleasant weather.*
    2. *Cirrostratus – Typically white, sheet-like clouds that cover the enter sky; they predict rain within twelve to twenty-four hours.*
    3. *Cirrocumulus – Typically long rows of small rounded puffs high in the sky usually seen in the winter, they predict fair, but cool weather.*

Alto Clouds
Information Learned
⇨ Alto clouds are mid-level clouds found between 6000 and 18,000 ft.
⇨ These clouds can be puffy or flat.
⇨ There are two types of alto clouds:
1. Altostratus – Typically greyish clouds that cover the sky with thinner patches where the sun is slightly visible; they form before a storm with continuous rain or snow.
2. Altocumulus – Typically groups of greyish, puffy clouds; they form on warm, humid mornings and predict a coming thunderstorm in the afternoon.

Stratus Clouds
Information Learned
⇨ Stratus clouds form low, flat layers under 6000 ft.
⇨ These clouds often block the sunshine.
⇨ There are three types of stratus clouds:
1. Stratus – Typically uniform grey clouds that cover the entire sky, light mist or rain can fall from these clouds.
2. Nimbostratus – Typically dark grey, wet looking clouds; these clouds bring on continuous rain or snow.
3. Stratocumulus – Typically rows of low, puffy, grey clouds with blue sky peeking out in between; rain rarely forms from these type of clouds.

Cumulus Clouds
Information Learned
⇨ Cumulus clouds are high, puffy clouds that can grow vertically between the layers.
⇨ There are two main types of cumulus clouds:
1. Cumulus – Typically white and puffy, sort of like cotton balls; these clouds are known as the fair weather clouds.
2. Cumulonimbus – Typically large puffy, towering clouds with grey flat bottoms and sometimes an anvil-shape at the top; these clouds are known as thunderstorm clouds and are associated with heavy rain, snow, hail, and lightning.

Vocabulary
Have the older students look up the following term in the glossary in the Appendix on pp. 109-110 or in a science encyclopedia. Then, have them copy the definition onto a blank index card or into their SCIDAT logbook.

✎ CLOUDS – A collection of condensed water vapor.

Scientific Demonstration: Cloud in a Bottle
Materials
☑ Hot water
☑ Glass jar with lid
☑ Crushed ice
☑ Match

Procedure
1. Fill the glass jar halfway with the hot water. (*Adults only. Use caution when touching the jar after this*

step as it will be warm.)

2. Then, flip the lid of the jar over, place it over the jar-opening, and fill it with crushed ice. Let the jar sit undisturbed for two to three minutes.
3. Light the match, throw it into the jar, and replace the lid. (*Adults only.*)
4. Wait about ten more minutes, watching the jar to observe of when a cloud begins to form in the jar. Once you see it, you can remove the lid and let the cloud escape while the students try to touch and catch it.

Explanation
The students should see a cloud form in the jar. Water vapor from the hot water rises in the jar and meets the cold air from the ice-filled lid, causing the water to condense. When the lighted match is added to the jar, it introduces tiny particles that the water vapor can condense around to form a cloud.

Take it Further
Have the students repeat the demonstration with you, once with adding no matches and once with adding two lit matches.

Multi-Week Projects and Activities

Multi-week Projects
- **Weather Poster** – Have the students add to their weather poster this week. (*Weather template pictures can be found in the Appendix on pg. 8.*)

Activities For This Week
- **Cloud Collage** – Have the students make a cloud collage depicting the three of the shapes of clouds - stratus, cumulus, and cirrus. You will need several cotton balls and a sheet of blue construction paper. Have the students use full cotton balls for cumulus clouds, which are large, puffy clouds. Then, have them stretch each cotton ball out a bit for stratus clouds, which are long, layered clouds. Finally, have them pull the cotton ball out until it is somewhat see-through for cirrus clouds, which are thin and wispy. Have the older students add a sentence describing each of these types of clouds.
- **Predict the Weather** – Have the students learn how the clouds can help them predict the weather using the information from the following website:
  - http://thehomeschoolscientist.com/using-clouds-to-predict-the-weather/
- **Cloud Video** – Have the students watch the following video about why clouds stay up:
  - https://www.youtube.com/watch?v=DjByja9ejTQ

Memorization

Copywork/Dictation
☞ **Copywork Sentence**

At high altitudes, it is typically much cooler because there is less air to trap the heat.

☞ **Dictation Selection**

Cirrus clouds are high and wispy and composed of ice. Alto clouds can be puffy or flat. Stratus clouds form low, flat layers that often block the sunshine. Cumulus clouds are high, puffy clouds that can grow vertically between the layers.

Notes

CHAPTER 12: BACK IN ALASKA

CHAPTER SUMMARY

The chapter opens with Blaine and Tracey zipping towards their next location – Alaska! When they get there, the twins find out that they will not be heading directly into Summer's Lab. Instead they will be competing in the annual P.B. and J., otherwise known as the Paintball Jamboree. Summer shares about the water cycle as they wait for the rest of the competitors. The three are soon joined by Ulysses, the robot squirrels, Skeeter and Tina (two local teachers), and Yotima, who rescued Tracey during their zoology leg. The first round of the game begins and we quickly learn that another more sinister player has entered the game: the Man with No Eyebrows. Tracey ends up capturing the "flag", or rather the information-filled tablet, which ends the first round of the game. The twins learn about fog before getting ready for the next round. The chapter ends with the Man with No Eyebrows ready to head into the fog to capture the twins!

SUPPLIES NEEDED

Demonstration	Projects and Activities
• Plastic baggie, Water, Tape	• Dry ice, A shallow container, Water

OPTIONAL SCHEDULE FOR TWO-DAYS-A-WEEK

Day 1	Day 2
☐ Read the section entitled "A Break in the Water Cycle" of Chapter 12 in *SSA Volume 4: Earth Science*.	☐ Read the section entitled "PB and J with a Side of Fog" of Chapter 12 in *SSA Volume 4: Earth Science*.
☐ Fill out an Earth Science Record Sheet on SL pg. 63 on the water cycle.	☐ Fill out an Earth Science Record Sheet on SL pg. 64 on fog.
☐ Add weather to the Weather Information Sheet on SL pg. 62.	☐ Add facts to the Climate Information on SL pg. 61; Add weather to the Weather Information Sheet on SL pg. 62.
☐ Do the demo entitled "Water Cycle in a Bag"; write information learned on SL pg. 67.	☐ Go over the vocabulary word and enter it into the Earth Science Glossary on SL pg. 99.
☐ Do the copywork or dictation assignment and add it to the Earth Science Notes on SL pg. 67.	☐ Work on one or all of the multi-week activities.

OPTIONAL SCHEDULE FOR FIVE-DAYS-A-WEEK

Day 1	Day 2	Day 3	Day 4	Day 5
☐ Read the section entitled "A Break in the Water Cycle" of Chapter 12 in *SSA Volume 4: Earth Science*.	☐ Read one or all of the assigned pages from the encyclopedia of your choice; write narration on the Earth Science Notes Sheet on SL pg. 67.	☐ Read the section entitled "PB and J with a Side of Fog" of Chapter 12 in *SSA Volume 4: Earth Science*.	☐ Read one of the additional library books.	☐ Do the copywork or dictation assignment and add it to the Earth Science Notes sheet on SL pg. 67.
☐ Fill out an Earth Science Record Sheet on SL pg. 63 on the water cycle.	☐ Do the demo entitled "Water Cycle in a Bag"; write information learned on SL pg. 67.	☐ Fill out an Earth Science Record Sheet on SL pg. 64 on fog.	☐ Go over the vocabulary word and enter it into the Earth Science Glossary on SL pg. 99.	☐ Add weather to the Weather Information Sheet on SL pg. 62.
☐ Add weather to the Weather Information Sheet on SL pg. 62.		☐ Add facts to the Climate Information on SL pg. 61.	☐ Choose one of the activities for the week to do; fill out the project record sheet on pg. 69.	☐ Work on one or all of the multi-week activities.

Science-Oriented Books

Living Book Spine
- Chapter 12 of *The Sassafras Science Adventures Volume 4: Earth Science*

Optional Encyclopedia Readings
- *Basher Science Planet Earth* pg. 80 (Water Cycle)
- *Usborne Children's Encyclopedia* pg. 14 (Section on the Water Cycle)
- *Discover Science Weather* pp. 20-21 (Blue planet)
- *Usborne Encyclopedia of Planet Earth* pg. 80 (Section on the Water Cycle)

Additional Living Books
- *The Water Cycle* by Helen Frost
- *The Water Cycle (Earth and Space Science)* by Craig Hammersmith
- *The Water Cycle (Water All Around)* by Rebecca Olien and Ted Williams
- *Fog and Mist (Watching the Weather)* by Elizabeth Miles
- *Fog (Weather)* by Helen Frost

Notebooking (SCIDAT Logbook Information)

This week, you can have the students begin to fill out the Climate Sheet for the Alaskan Boreal Forest. They can also fill out the first part of their weather record sheets and the logbook sheets for the water cycle and fog. Here is the information they could include:

Climate Sheets
Area Map – Have the students color the region where the boreal forest are found in Alaska. (*See map for answers.*)

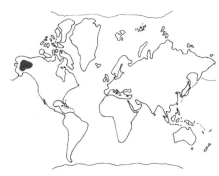

Climate Information – This week, the students could include the following:
- ⇨ The boreal forest receives mostly snowfall for its precipitation. It does rain during the summer months, but the ground, rivers, and lakes get a significant amount of their water from snowmelt.
- ⇨ The temperature changes drastically from summer to winter in the boreal forest. In the summer, the temperature can range from 30°F to 70°F. In the winter, the temperature can range from -65°F to 30°F.

Interesting Facts – Answers will vary. Students can add the following fact:
- ⇨ The boreal coniferous forest is also known as the taiga. It is the world's largest biome.
- ⇨ Because of the range of temperatures, the growing season in the taiga is very short and many animals migrate in and out of the forest.

Other Types – This week, the students could include the following:
- ⇨ The name boreal comes from the Greek god of the North Wind, who was name Boreas.

Weather Record Sheets
Have the students record the weather from your area or from Alaska over the week.

Earth Science Record Sheets
Water Cycle
Information Learned
- ⇨ The water on Earth is limited.

⇨ It regularly changes form in a process we call the water cycle.
⇨ The process includes evaporation, condensation, precipitation, and collection.
⇨ The Steps of the Water Cycle:
   1. The sun's heat evaporates water from the world's oceans, lakes, and rivers.
   2. The water rises into the air as water vapor.
   3. Water vapor condenses to form clouds.
   4. As the water vapor cools, the clouds become heavy and water falls as rain or snow.
   5. Rainwater and snow melt flow back into the rivers, lakes, and oceans.

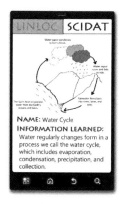

Fog
Information Learned
⇨ Fog is a thick cloud of tiny water droplets that are suspended very close to the Earth's surface. In other words, fog is a cloud on the ground.
⇨ Fog forms in the same way that clouds do. The water vapor in the air close to the ground condenses and forms a cloud.
⇨ Fog typically only happens when it is very humid, or full of water vapor, as there has to be a lot of water vapor in the air for fog to form.
⇨ There are several different types of fog based on how it forms.
   1. Radiation fog – This fog forms as humid air near the ground cools overnight, so this type of fog is typically seen in the morning and "burns" off as the sun starts shining.
   2. Advection fog – This fog forms when warm, moist tropical air moves over a cooler surface. It typically forms on the coast.
   3. Valley fog – This fog forms when moist air is trapped by the mountains in a valley.

Vocabulary
Have the older students look up the following term in the glossary in the Appendix on pp. 109-110 or in a science encyclopedia. Then, have them copy the definition onto a blank index card or into their SCIDAT logbook.

📋 FOG – A thick cloud of tiny water droplets that are suspended very close to the Earth's surface.

Scientific Demonstration: Water Cycle in a Bag
Materials
☑ Plastic baggie
☑ Water
☑ Tape

Procedure
1. Have the students add a cup of water to the plastic baggie.
2. Blow a bit of air into the baggie and seal it up tight.
3. Find a window in direct sunlight and tape the baggie to the window.
4. Have the students observe what happens in the baggie.

Explanation
The purpose of this demonstration is to allow the students to see and record the pressure changes found in their area. If the students also observe the weather, they should see that the higher the pressure, the clearer the weather.

Take It Further
Use a permanent marker to draw the steps of the water cycle evaporation, condensation,

precipitation, and collection. The students can include sun, clouds, land, and bodies of water.

Multi-Week Projects and Activities

Multi-week Projects
- **WEATHER POSTER** – Have the students add to their weather poster this week. (*Weather template pictures can be found in the Appendix on pg. 8.*)

Activities For This Week
- **WATER CYCLE** – Have the students view the following interactive water cycle:
  - http://player.discoveryeducation.com/views/hhView.cfm?guidAssetId=087777c8-4ff0-45d2-878f-e7cd90f7ee19
- **WATER CYCLE POSTER** – Have the students make a poster depicting the steps of the water cycle. You can use the template found on pp. 100-101 of the Appendix, or you can have the students draw their own.
- **FOG** – Create some homemade fog with dry ice, a shallow container, and water. Add the dry ice to the bottom of the container, slowly pour water over it, and observe the fog that is created.

Memorization

Copywork/Dictation

- **COPYWORK SENTENCE**
  Fog is a thick cloud of tiny water droplets that are found close to the Earth's surface.

- **DICTATION SELECTION**
  In the water cycle, the sun's heat evaporates water from the world's oceans, lakes, and rivers. Then, the water rises into the air as water vapor. Next, water vapor condenses to form clouds. As the water vapor cools, the clouds become heavy and water falls as rain or snow. And finally, rainwater and snow melt flow back into the rivers, lakes, and oceans.

Notes

Chapter 13: The Forget-o-nator

Chapter Summary

The chapter opens with a series of flashes between the different players of the P.B. and J. games. Paintballs fly, Blaine is hit, but in the end Tracey captures the tablet once more! Summer doesn't immediately appear, so Skeeter and Tina step in to share about the nitrogen cycle with the twins. When the expert/scientist still doesn't show, the group heads back to her lab to see if she is there. They can't find her there either, so Skeeter and Tina finishes sharing about the natural cycles with the twins. Meanwhile, we learn that Summer has been captured by the Man with No Eyebrows and that he plans to put her in the Forget-O-Nator. Summer recognizes the Man with No Eyebrows as her former classmate, Thaddeus. Back at the lab, Tracey calls Uncle Cecil, who encourages them to go ahead and zip onto the next location in the Pacific Ocean. The chapter ends with Summer falling to the floor in the Forget-O-Nator, completely devoid of memory.

Supplies Needed

Demonstration	Projects and Activities
• Soil sample, Coffee filter, Rubber band • 2 Cups, Distilled water, Aquarium test strip	• No Additional Supplies Needed

Optional Schedule for Two-Days-A-Week

Day 1	Day 2
☐ Read the section entitled "Competing Cycles" of Chapter 13 in *SSA Volume 4: Earth Science*. ☐ Fill out an Earth Science Record Sheet on SL pg. 65 on the nitrogen cycle. ☐ Add facts to the Climate Information on SL pg. 61; ☐ Add weather to the Weather Information Sheet on SL pg. 62. ☐ Do the copywork or dictation assignment and add it to the Earth Science Notes on SL pg. 68.	☐ Read the section entitled "Phosphorus Predicaments" of Chapter 13 in *SSA Volume 4: Earth Science*. ☐ Fill out an Earth Science Record Sheet on SL pg. 66 on the phosphorus cycle. ☐ Add weather to the Weather Information Sheet on SL pg. 62; Go over the vocabulary word and enter it into the Earth Science Glossary on SL pg. 99. ☐ Do the demo entitled "Soil Test"; write information learned on SL pg. 68. ☐ Work on one or all of the multi-week activities.

Optional Schedule for Five-Days-A-Week

Day 1	Day 2	Day 3	Day 4	Day 5
☐ Read the section entitled "Competing Cycles" of Chapter 13 in *SSA Volume 4: Earth Science*. ☐ Fill out an Earth Science Record Sheet on SL pg. 65 on the nitrogen cycle. ☐ Add facts to the Climate Information on SL pg. 61.	☐ Read the section entitled "Phosphorus Predicaments" of Chapter 13 in *SSA Volume 4: Earth Science*. ☐ Fill out an Earth Science Record Sheet on SL pg. 66 on the phosphorus cycle. ☐ Add weather to the Weather Information Sheet on SL pg. 62.	☐ Read one or all of the assigned pages from the encyclopedia of your choice; write narration on the Earth Science Notes Sheet on SL pg. 68. ☐ Do the demo entitled "Soil Test"; write information learned on SL pg. 68.	☐ Read one of the additional library books. ☐ Go over the vocabulary word and enter it into the Earth Science Glossary on SL pg. 99. ☐ Choose one of the activities for the week to do; fill out the project record sheet on pg. 70.	☐ Do the copywork or dictation assignment and add it to the Earth Science Notes sheet on SL pg. 68. ☐ Add weather to the Weather Information Sheet on SL pg. 62. ☐ Work on one or all of the multi-week activities.

Science-Oriented Books

Living Book Spine
📖 Chapter 13 of *The Sassafras Science Adventures Volume 4: Earth Science*

Optional Encyclopedia Readings
- *Basher Science Planet Earth* (No pages scheduled.)
- *Discover Science Weather* (No pages scheduled.)
- *Usborne Children's Encyclopedia* (No pages scheduled.)
- *Usborne Encyclopedia of Planet Earth* pp. 52-53 (Natural Cycles)

Additional Living Books
📖 *The Nitrogen Cycle (Cycles in Nature)* by Suzanne Slade

Notebooking (SCIDAT Logbook Information)

This week, you can have the students fill out the second part of their weather record sheets and the logbook sheets for the nitrogen and phosphorus cycles. Here is the information they could include:

Climate Sheets
There is no information for the students to add to the climate sheets for this week.

Weather Record Sheets
Have the students record the weather from your area or from Alaska over the week.

Earth Science Record Sheets

Nitrogen Cycle
Information Learned
- Nitrogen is an essential elemental for life on earth. The nitrogen cycle explains how this element moves between plants, animals, bacteria, the air, and the soil.
- Nitrogen gas is found in the atmosphere; approximately 78% of our atmosphere is nitrogen. Lightning combines nitrogen and oxygen into nitrogen-containing compounds, which then fall to the ground in the rain.
- The nitrogen-containing compounds soak into the soil. Bacteria covert those nitrogen compounds in the soil to make it usable for plants.
- Plants take up converted nitrogen compounds from the soil and animals get the nitrogen they need from eating these plants. When the plants and animals die, they are broken down by bacteria known as decomposers. This process puts usable nitrogen-containing compounds back into the soil to begin the cycle again.
- The excess nitrogen-containing compounds are absorbed by special bacteria that convert the compounds and then release nitrogen into the air once more.

Phosphorus Cycle
Information Learned
- Phosphorus is another essential element for life.
- The phosphorus cycle explains how this element moves between plants, animals, water, and the soil, but it is not a true cycle because a fair amount of phosphorus is lost along the way.
- Rock found in the ground contains phosphate, a phosphorus-containing

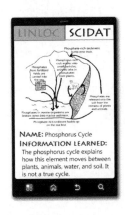

compound. When it rains, some of this phosphate leaks out of the rock into the surrounding soil and water.
- ⇨ *Plants absorb this phosphate and animals can get the phosphorus they need by eating these plants or by drinking phosphate-rich water.*
- ⇨ *When the plants and animals die, decomposers break them down and release phosphates into the surrounding soil.*
- ⇨ *In the ocean, much of the phosphates fall to the bottom, rendering them inaccessible to plants, which means that they cannot be used to continue the cycle. This accounts for the most of the "lost" phosphorus in the cycle.*

Vocabulary

Have the older students look up the following term in the glossary in the Appendix on pp. 109-110 or in a science encyclopedia. Then, have them copy the definition onto a blank index card or into their SCIDAT logbook.

- **NATURAL CYCLE** – A naturally occurring event that happens over and over again in the same order.

Scientific Demonstration: Soil Test

Materials
- ☑ Soil sample
- ☑ Coffee filter
- ☑ Rubber band
- ☑ 2 Cups
- ☑ Distilled water
- ☑ Aquarium test strip (one that tests the pH and nitrate levels)

Procedure
1. Have the students determine an area in your back yard to collect a soil sample. Clear the area from debris and dig down about four to six inches. Collect about a cup of soil from the sample area.
2. Head inside to dry out the sample. You can do this by spreading it out on a cookie sheet and letting it sit overnight on the counter, or by placing the sheet in a oven on low heat for several hours.
3. Once the soil is dry, break it apart with your hands until you have a fine powder. In one of the cups, add ½ cup of the soil powder to ½ cup of distilled water. Mix well.
4. Place the coffee filter over the second cup, creating a pocket, and use the rubber band to secure it in place. Pour the soil sample mixture from the first cup into the second to filter it out.
5. Once the water has been filtered, dip the aquarium test strip into the water and wait the prescribed amount of time for the results to develop. Use the scale on the test strip's container to determine your results.

Explanation
The results will vary based on the condition of your soil. Generally, the students should see that soils that are acidic, meaning the dirt has a pH level of less than 7, are lacking in nitrogen. In other words, if the pH level of the sample is low, the nitrate level will also be low.

Take it Further
Have the students repeat the demonstration with soil from different depths: one sample that is shallower and one that is deeper.

Multi-Week Projects and Activities

Multi-week Projects

✂ **Weather Poster** – Have the students add to their weather poster this week. (*Weather template pictures can be found in the Appendix on pg. 8.*)

Activities For This Week

✂ **Videos** – Have the students learn more about the two natural cycles from this week by watching the following videos:
- Nitrogen Cycle: https://www.youtube.com/watch?v=ZaFVfHftzpI
- Phosphorus Cycle: https://www.youtube.com/watch?v=wdAzQSuypCk

✂ **Nitrogen cycle poster** – Have the students make a poster depicting the steps of the nitrogen cycle. You can use the template found on pp. 102-103 of the Appendix, or you can have the students draw their own.

✂ **Phosphorus cycle poster** – Have the students make a poster depicting the steps of the phosphorus cycle. You can use the template found on pp. 104-105 of the Appendix, or you can have the students draw their own.

Memorization

Copywork/Dictation

☞ **Copywork Sentence**

A natural cycle is an event that happens over and over again in the same order in nature.

☞ **Dictation Selection**

Nitrogen is an essential elemental for life on earth. The nitrogen cycle explains how this element moves between plants, animals, bacteria, the air, and the soil. Phosphorus is another essential element for life. The phosphorus cycle explains how this element moves between plants, animals, water, and the soil. It is not a true cycle because a fair amount of phosphorus is lost along the way.

Notes

Chapter 14: A Watery Landing... Again

Chapter Summary

The chapter opens with the Sassafras twins landing in water, again. They are quickly rescued by a floating bunk bed, which they ride past a coral reef and onto the Western Garbage Patch. Once there, the "meet" several people made out of trash created by their local expert, Billfrey Battabingo. He turns out to be a bit off his rocker, but he does manage to tell the twins about the coral reef they saw, in between introducing his recycled-friends and acting like a bird. Blaine and Tracey learn a bit more about the garbage patch and the currents found in the ocean before going fishing with their expert. They catch a few fish and cook them up for dinner. As they finish their meal, they hear a sudden roar coming from a box labeled "Taipei Zoo." Within moments, a bear and several baboons begin chasing the three garbage-patch dwellers. The chapter ends with Blaine and Tracey finding themselves mixed up in a hilarious, but frightening, chase!

Supplies Needed

Demonstration	Projects and Activities
• Water, Cup, Ice, Bowl, Blue food coloring	• Plastic bottle, Water, Blue food coloring, Oil, Duct tape • Coral sample

Optional Schedule for Two-Days-A-Week

Day 1	Day 2
☐ Read the section entitled "Careening towards Coral" of Chapter 14 in *SSA Volume 4: Earth Science*. ☐ Fill out an Earth Science Record Sheet on SL pg. 73 on the coral reef. ☐ Add weather to the Weather Information Sheet on SL pg. 72. ☐ Do the demo entitled "Moving Currents"; write information learned on SL pg. 77. ☐ Do the copywork or dictation assignment and add it to the Earth Science Notes on SL pg. 77.	☐ Read the section entitled "Drifting in Currents..." of Chapter 14 in *SSA Volume 4: Earth Science*. ☐ Fill out an Earth Science Record Sheet on SL pg. 74 on currents. ☐ Add facts to the Climate Information on SL pg. 71; Add weather to the Weather Information Sheet on SL pg. 72. ☐ Go over the vocabulary words and enter them into the Earth Science Glossary on SL pp. 99-100. ☐ Work on one or all of the multi-week activities.

Optional Schedule for Five-Days-A-Week

Day 1	Day 2	Day 3	Day 4	Day 5
☐ Read the section entitled "Careening towards Coral" of Chapter 14 in *SSA Volume 4: Earth Science*. ☐ Fill out an Earth Science Record Sheet on SL pg. 73 on the coral reef. ☐ Add weather to the Weather Information Sheet on SL pg. 72.	☐ Read the section entitled "Drifting in Currents..." of Chapter 14 in *SSA Volume 4: Earth Science*. ☐ Fill out an Earth Science Record Sheet on SL pg. 74 on currents. ☐ Add facts to the Climate Information on SL pg. 71.	☐ Read one or all of the assigned pages from the encyclopedia of your choice; write narration on the Earth Science Notes Sheet on SL pg. 77. ☐ Do the demo entitled "Moving Currents"; write information learned on SL pg. 77.	☐ Read one of the additional library books. ☐ Go over the vocabulary words and enter them into the Earth Science Glossary on SL pp. 99-100. ☐ Choose one of the activities for the week to do; fill out the project record sheet on pg. 79.	☐ Do the copywork or dictation assignment and add it to the Earth Science Notes sheet on SL pg. 77. ☐ Add weather to the Weather Information Sheet on SL pg. 72. ☐ Work on one or all of the multi-week activities.

Science-Oriented Books

Living Book Spine
- Chapter 14 of *The Sassafras Science Adventures Volume 4: Earth Science*

Optional Encyclopedia Readings
- *Basher Science Planet Earth* pp. 116-117 (Open Ocean), pp. 120-121 (Coral Reef)
- *Usborne Children's Encyclopedia* pp. 38-39 (Currents), pp. 82-83 (Coral Reefs)
- *Discover Science Weather* (No pages scheduled.)
- *Usborne Encyclopedia of Planet Earth* pp. 50-51 (Air and Ocean Currents)

Additional Living Books
- *National Geographic Readers: Coral Reefs* by Kristin Rattini
- *Exploring Coral Reefs* by Anita Ganeri
- *Coral Reefs* by Seymour Simon
- *Over in the Ocean: In a Coral Reef* by Marianne Berkes

Notebooking (SCIDAT Logbook Information)

This week, you can have the students begin to fill out the Climate Sheet for the Pacific Ocean. They can also fill out the first part of their weather record sheets and the logbook sheets for coral reefs and currents. Here is the information they could include:

Climate Sheets
Area Map – Have the students color the region where the Pacific Ocean is found. (*See map for answers.*)

Interesting Facts – Answers will vary. Students can add the following facts:
⇨ The Western Garbage Patch is an area of spinning debris, or trash.
⇨ It was created by the currents that flow through the ocean and the discarded non-biodegradable trash: things like broken-up plastic bottles and rubber shoes.
⇨ The currents bring the debris to a stable location known as a gyre, where the trash is trapped. It spins slowly round and round, but doesn't move out of the area, which creates a vortex.
⇨ Some of the debris remains on the surface, but about 70% is just below or closer to the ocean floor.

Weather Record Sheets
Have the students record the weather from your area or from area where the Western Garbage Patch is found over the week.

Earth Science Record Sheets
Coral Reef
Information Learned
⇨ Coral reefs are underwater structures made from the skeletons of tiny animals known as corals.
⇨ The corals grow in colonies, as older corals die off, new ones grow on top of them. Over time, the layers build upon each other to create a large structure.
⇨ Many different varieties of coral growing and building together form a reef,

literally teeming with life.

⇨ *Coral reefs are home to the greatest mix of plants and animals found in the oceans. They are home to more than fifteen percent of all fist species, but only cover one percent of the Earth's surface.*

⇨ *Reefs are found throughout the warm tropical waters of the oceans, but the largest coral reef is the Great Barrier Reef just off the coast of Australia, which stretches for over fourteen hundred miles.*

⇨ *Coral reefs form in shallow waters of the world's oceans, with plenty of wave action. The waves do not allow sediment to fall on the reef and the waves bring nutrients and minerals to the animals living in the reef.*

Currents
Information Learned

⇨ The water in the ocean is constantly moving due to currents.

⇨ There are two main types of currents – surface and deep:
1. Surface currents are due to the winds that affect the given area and usually affect the top of the ocean. These currents generally push water towards land and are responsible for the waves we see on the beach.
2. Deep currents are due to the sinking action of cold water that comes from the poles, which then drifts to the equator, warms up, and rises to the surface again and drifts towards the poles where it cools off, creating a cycle of rising and sinking water throughout the ocean. Deep water currents are also affected by the salinity, or saltiness, of the water.

⇨ Currents in the ocean are also produced by the spinning motion of the Earth, which is part of the Coriolis Effect.

Vocabulary

Have the older students look up the following terms in the glossary in the Appendix on pp. 109-110 or in a science encyclopedia. Then, have them copy each definition onto a blank index card or into their SCIDAT logbook.

- **CORAL** – A group of tiny animals with a hard shells, similar to rock; they live together in a colony that builds upon itself.
- **CURRENTS** – The movement of water or air in a particular direction, usually due to a difference in temperature

Scientific Demonstration: Moving Currents
Materials
☑ Water
☑ Cup
☑ Ice
☑ Bowl
☑ Blue food coloring

Procedure
1. Add the ice cubes plus enough cold water to cover them to the bowl. Gently place the food coloring in the ice water to cool it off.
2. Next, add 1 cup of warm water to the cup.
3. After ten minutes, pull the food coloring out of the ice bath and add several drops to the cup. Have the students observe what happens.

Explanation

The purpose of this demonstration is to allow the students to see how the difference in temperature creates the movement of currents. When the ice cold food coloring meets the warm water, it instantly starts twirling and swirling through the water. The same movement is created on a larger scale in the ocean, which forms the currents.

TAKE IT FURTHER
Have the students repeat the demonstration, only this time use ice cold food coloring and ice cold water. (*The students should see that the mixing is much slower than before.*)

MULTI-WEEK PROJECTS AND ACTIVITIES

MULTI-WEEK PROJECTS
- **WEATHER POSTER** – Have the students add to their weather poster this week. (*Weather template pictures can be found in the Appendix on pg. 8.*)

ACTIVITIES FOR THIS WEEK
- **CORAL REEF** – Have the students learn more about the coral reef by watching the following video:
  - https://www.youtube.com/watch?v=J2BKd5e15Jc
- **WAVES IN A BOTTLE** – Have the students create visible waves in a bottle! You will need a plastic water bottle, water, blue food coloring, oil (vegetable or mineral), and duct tape. Fill the bottle a third of the way with water and add a few drops of blue food coloring. Swirl the bottle around to mix well. Then, slowly add an equal amount of oil to the bottle. Seal the bottle and use the duct tape to make sure no water or oil escape from the top. Tip the bottle over on its side and gently sway it back and forth to create a wave motion.
- **MICROSCOPE WORK** – Have the students look at coral under the microscope. You will need to obtain a sample of dried coral, which means that the students will see the hard shell structure of the colony, but not the live animals. Have the students complete one of the microscope worksheets found on pp. 97-98 of the Appendix. If you do not have a microscope, you can view several images of coral under a microscope here:
  - http://www.arkive.org/montastrea-coral/montastrea-serageldini/image-G71339.html

MEMORIZATION

COPYWORK/DICTATION
- **COPYWORK SENTENCE**
  The water in the ocean is always moving due to currents.
- **DICTATION SELECTION**
  There are two main types of currents: surface currents and deep currents. Surface currents are caused by the winds that blow over an area of the ocean. These currents push water towards land and are responsible for the waves on the beach. Deep currents are caused by the sinking action of cold water that comes from the poles and drifts to the equator. When the cold water warms up, it rises to the surface again and drifts towards the poles, where it cools off. This creates a cycle of rising and sinking water throughout the ocean.

NOTES

Chapter 15: The Threat of Thaddaeus

Chapter Summary

The chapter opens with Uncle Cecil learning about Typhoon Thaddaeus, which is bearing down on where the twins are located. We flash back to the garbage patch, where Blaine and Tracey have managed to re-capture the animals, but not before the chase has caused that part of the garbage patch to begin to break up. They quickly realize that their backpacks, with their phones in them, are on one of the pieces that has broken off and is floating away. Billfrey's trash-friends are also floating away as well. The three work together to get the pieces back together as a raft and in the process the twins learn about the world's oceans. As their raft drifts through the waters, the winds pick up and sharks circle. Even so, the twins are able to fall asleep only to be awakened by Billfrey's declaration that a typhoon is approaching. After they learn about these types of storms, the twins spot a ship on the horizon. It turns out to be a rescue boat manned by none other than the P.R.O. pirates and their captain, Peace Beard. The newly-minted pira-medics rescue Billfrey and his recycled companions, but they apologize because they run out of room on the boat, as they are only certified to rescue six people at a time. The chapter ends with Blaine and Tracey watching the ship sail off, leaving them floating through the ocean and heading straight for the typhoon.

Supplies Needed

Demonstration	Projects and Activities
• 2 Eggs, 2 Tall cups, Water, Salt	• Corn syrup, Dish soap, Water, Oil, Rubbing alcohol, Black, purple, and blue food coloring • Plastic bottle, Opaque liquid soap, Blue food coloring, Water, Duct tape

Optional Schedule for Two-Days-A-Week

Day 1	Day 2
☐ Read the section entitled "Oceanic Occurrences" of Chapter 15 in *SSA Volume 4: Earth Science*.	☐ Read the section entitled "Typhoon Thaddaeus" of Chapter 15 in *SSA Volume 4: Earth Science*.
☐ Fill out an Earth Science Record Sheet on SL pg. 75 on the oceans.	☐ Fill out an Earth Science Record Sheet on SL pg. 76 on hurricanes.
☐ Add weather to the Weather Information Sheet on SL pg. 72; Work on one or all of the multi-week activities.	☐ Add facts to the Climate Information on SL pg. 71; Add weather to the Weather Information Sheet on SL pg. 72.
☐ Do the demo entitled "Ocean Float"; write information learned on SL pg. 78.	☐ Do the copywork or dictation assignment and add it to the Earth Science Notes on SL pg. 78.

Optional Schedule for Five-Days-A-Week

Day 1	Day 2	Day 3	Day 4	Day 5
☐ Read the section entitled "Oceanic Occurrences" of Chapter 15 in *SSA Volume 4: Earth Science*. ☐ Fill out an Earth Science Record Sheet on SL pg. 75 on the oceans. ☐ Add weather to the Weather Information Sheet on SL pg. 72.	☐ Read one or all of the assigned pages from the encyclopedia of your choice; write narration on the Earth Science Notes Sheet on SL pg. 78. ☐ Do the demo entitled "Water Cycle in a Bag"; write information learned on SL pg. 78.	☐ Read the section entitled "Typhoon Thaddaeus" of Chapter 15 in *SSA Volume 4: Earth Science*. ☐ Fill out an Earth Science Record Sheet on SL pg. 76 on hurricanes. ☐ Add facts to the Climate Information on SL pg. 71.	☐ Read one of the additional library books. ☐ Choose one of the activities for the week to do; fill out the project record sheet on pg. 80. ☐ Add weather to the Weather Information Sheet on SL pg. 72.	☐ Do the copywork or dictation assignment and add it to the Earth Science Notes sheet on SL pg. 78. ☐ Work on one or all of the multi-week activities.

Science-Oriented Books

Living Book Spine
- Chapter 15 of *The Sassafras Science Adventures Volume 4: Earth Science*

Optional Encyclopedia Readings
- *Basher Science Planet Earth* pp. 60-61 (Ocean) pp. 88-89 (Hurricane)
- *Discover Science Weather* pg. 27 (Hurricane Damage)
- *Usborne Children's Encyclopedia* pp. 34-35 (Seas and Oceans)
- *Usborne Encyclopedia of Planet Earth* pp. 134-135 (Seas and Oceans)

Additional Living Books
- *Eye Wonder: Oceans* (DK Eyewonder)
- *Basher Science: Oceans: Making Waves!*
- *Hurricane Watch (Let's-Read-and-Find-Out Science 2)* by Melissa Stewart and Taia Morley
- *Hurricanes (Little Scientist)* by Martha E. H. Rustad
- *The Magic School Bus Inside A Hurricane* by Joanna Cole

Notebooking (SCIDAT Logbook Information)

This week, you can have the students fill out the second part of their weather record sheets and the logbook sheets for the oceans and hurricanes. Here is the information they could include:

Climate Sheets
Interesting Facts – Answers will vary. Students can add the following facts:
⇨ Earth has three main types of water:
1. Seawater, which has a higher concentration of salt, is typically found in the world's oceans.
2. Freshwater, which has little to no salt, is typically found in rivers, lakes, ponds, streams, and wetlands.
3. Brackish water, which has less salt than saltwater, but more than seawater, is typically found where a freshwater source meets the ocean, such as in a delta and estuaries. The Baltic Sea is the world's largest brackish water source

⇨ A sea is much smaller than an ocean and is typically partially enclosed by land. Seas are usually found where oceans and land meet.

Weather Record Sheets
Have the students record the weather from your area or from area where the Western Garbage Patch is found over the week.

Earth Science Record Sheets
Oceans
Information Learned
⇨ Oceans cover nearly two-thirds of the Earth's surface.
⇨ There are five main oceans:
1. The Pacific – the largest ocean, covers nearly a third of Earth's surface, located between North and South America and Asia and Australia.
2. The Atlantic – the second largest ocean, is located between North and South

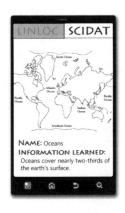

American and Europe and Africa.
  3. *The Arctic – located near the North Pole.*
  4. *The Southern – located near the South Pole.*
  5. *The Indian – located between Africa, Asia, and Australia.*
- ⇨ *All the world's oceans are joined up, meaning that the currents can carry water from ocean to ocean.*
- ⇨ *The floor of the ocean is similar to Earth's surface – it has valleys, mountains, hills, and trenches.*
- ⇨ *In general, the seabed slopes downhill gradually, away from land, to form a large shelf known as the continental shelf. It then drops quickly away to the deeper part of the ocean at the continental slope. The deepest part of the ocean is called the abyssal plain.*
- ⇨ *The deepest part of the ocean is the Marianas Trench, which is located in the Pacific Ocean off the coast of the Philippines.*

Hurricanes
Information Learned

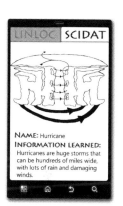

- ⇨ *Hurricanes are huge storms with lots of rain and damaging winds. They can be hundreds of miles wide.*
- ⇨ *When they hit land, they can cause lots of damage and flooding. However, when they hit land, hurricanes begin to weaken.*
- ⇨ *Hurricanes only form in warm, wet conditions, like those found in the tropics. For a hurricane to form, the ocean water needs to be 80°F or warmer.*
- ⇨ *The winds spiral around, creating an eye at the center where things are calm. In the northern hemisphere, they rotate counter-clockwise; in the southern hemisphere, they typically rotate clockwise.*
- ⇨ *Hurricanes typically form during the hurricane season, which is June 1 to November 30.*
- ⇨ *In the Atlantic Ocean, Gulf of Mexico, or Eastern Pacific Ocean, these storms are known as hurricanes.*
- ⇨ *In the Western Pacific Ocean, they are known as typhoons.*
- ⇨ *In the Indian Ocean, Bay of Bengal, or near Australia, these storms are known as cyclones.*

Vocabulary

There are no vocabulary words for this chapter.

Scientific Demonstration: Ocean Float

Materials
- ☑ 2 Eggs
- ☑ 2 Tall cups
- ☑ Water
- ☑ Salt

Procedure
1. Have the students add about a cup and a half of warm water to each cup. Label the first cup freshwater and the second saltwater.
2. To the second cup, have the students add 5 to 6 tablespoons of salt. Stir until almost all of the salt dissolves to mix up some ocean saltwater.
3. Next, have the students gently place an egg in each cup and observe what happens.

Explanation

The purpose of this demonstration is to allow the students to see how freshwater and saltwater (like that in the ocean) differ. The addition of salt causes the water to be more dense, making it easier for objects to float in the saltwater.

TAKE IT FURTHER
   Have the students repeat the demonstration will different objects they find around the house, such as pencils, lego mini-figs, and pieces of trash.

Multi-Week Projects and Activities

Multi-week Projects
✂ **WEATHER POSTER** – Have the students add to their weather poster this week. (*Weather template pictures can be found in the Appendix on pg. 8.*)

Activities For This Week
✂ **LAYERS OF THE OCEAN** – Teach the students about the layers of the ocean by creating a liquid representation of the layers in a bottle. You will need corn syrup, dish soap, water, oil, rubbing alcohol, and black, purple, and blue food coloring. The directions for this project can be found in the following post:
   🖱 http://www.icanteachmychild.com/make-ocean-zones-jar/

✂ **OCEANS OF THE WORLD** – Have the students label and color the world's major oceans (Atlantic, Pacific, Arctic, Southern, and Indian Oceans) on a map of the earth. You can use the map template provided in the Appendix on pg. 106. Here are the answers for your convenience:

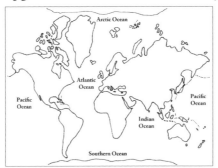

✂ **HURRICANE IN A BOTTLE** – Have the students recreate the swirling action of a hurricane in a bottle! You will need a plastic water bottle, opaque liquid soap that contains glycol stearate (such as the Softsoap brand), blue food coloring, water, and duct tape. Fill the bottle a third of the way with the liquid soap and add a few drops of the food coloring. Then, slowly add warm water until the liquid reaches the top. If there are any bubbles, add enough water to force them to go up and over the top. Seal the bottle and use the duct tape to make sure no liquid escapes from the top. Tilt the bottle back and forth to recreate the swirling motion of the winds in a hurricane.

Memorization

Copywork/Dictation
☞ **COPYWORK SENTENCE**
   Hurricanes are huge storms with lots of rain and wind.

☞ **DICTATION SELECTION**
   Hurricanes are huge storms with lots of rain and damaging wind. They can be hundreds of miles wide. In the Atlantic Ocean, Gulf of Mexico, or Eastern Pacific Ocean, these storms are known as hurricanes. In the Western Pacific Ocean, they are known as typhoons. In the Indian Ocean, Bay of Bengal, or near Australia these storms are known as cyclones.

Notes

Chapter 16: Quick! To Switzerland!

Chapter Summary

The chapter opens with the twins adrift in a storm-tossed ocean. Thankfully, they are soon rescued by Yotimo, Skeeter, and Tina in Summer's heliquickter. They had been sent by Uncle Cecil when he could not reach the twins. Yotimo flies the twins out of danger and they quickly zip off to the next location in Switzerland. The twins find themselves being interrogated by the SSS, the Swiss Secret Service, about their involvement with Yuroslav Boganovich. The agents quickly realize that the twins are not on Yuroslav's side and the twins agree to help the agents capture him. The twins' local expert, Evan DeBlose, tells them about groundwater and the mission as they prepare to head out. Blaine and Tracey learn that they will sit in a van to help identify Yuroslav before an undercover agent pretends to make a deal with him and takes him down. The deal will go down at the local kayaking competition, which Evan tells them about, along with sharing more information on waterfalls. The chapter ends with Tracey spotting Yuroslav.

Supplies Needed

Demonstration	Projects and Activities
• Plastic bottle, Cotton balls, Gravel, Sand, Soil • Duct tape, Water	• Sponge, Bar of soap, like Ivory

Optional Schedule for Two-Days-A-Week

Day 1	Day 2
☐ Read the section entitled "Grim Groundwater" of Chapter 16 in *SSA Volume 4: Earth Science*. ☐ Fill out an Earth Science Record Sheet on SL pg. 83 on groundwater. ☐ Add weather to the Weather Information Sheet on SL pg. 82. ☐ Do the demo entitled "Groundwater Filter"; write information learned on SL pg. 87. ☐ Do the copywork or dictation assignment and add it to the Earth Science Notes on SL pg. 87.	☐ Read the section entitled "I Spy: Waterfall" of Chapter 16 in *SSA Volume 4: Earth Science*. ☐ Fill out an Earth Science Record Sheet on SL pg. 84 on waterfalls. ☐ Add facts to the Climate Information on SL pg. 81; Add weather to the Weather Information Sheet on SL pg. 82. ☐ Go over the vocabulary word and enter it into the Earth Science Glossary on SL pg. 100. ☐ Work on one or all of the multi-week activities.

Optional Schedule for Five-Days-A-Week

Day 1	Day 2	Day 3	Day 4	Day 5
☐ Read the section entitled "Grim Groundwater" of Chapter 16 in *SSA Volume 4: Earth Science*. ☐ Fill out an Earth Science Record Sheet on SL pg. 83 on groundwater. ☐ Add weather to the Weather Information Sheet on SL pg. 82.	☐ Read one or all of the assigned pages from the encyclopedia of your choice; write narration on the Earth Science Notes Sheet on SL pg. 87. ☐ Do the demo entitled "Groundwater Filter"; write information learned on SL pg. 87.	☐ Read the section entitled "I Spy: Waterfall" of Chapter 16 in *SSA Volume 4: Earth Science*. ☐ Fill out an Earth Science Record Sheet on SL pg. 84 on waterfalls. ☐ Add facts to the Climate Information on SL pg. 81.	☐ Read one of the additional library books. ☐ Go over the vocabulary word and enter it into the Earth Science Glossary on SL pg. 100. ☐ Choose one of the activities for the week to do; fill out the project record sheet on pg. 89.	☐ Do the copywork or dictation assignment and add it to the Earth Science Notes sheet on SL pg. 87. ☐ Add weather to the Weather Information Sheet on SL pg. 82. ☐ Work on one or all of the multi-week activities.

Science-Oriented Books

Living Book Spine
- Chapter 16 of *The Sassafras Science Adventures Volume 4: Earth Science*

Optional Encyclopedia Readings
- *Basher Science Planet Earth* pp. 108-109 (Temperate Forest)
- *Usborne Children's Encyclopedia* pg. 25 (Section on Waterfalls)
- *Discover Science Weather* (No pages scheduled.)
- *Usborne Encyclopedia of Planet Earth* pp. 128-129 (Water in the Ground), pg. 125 (Section on Waterfalls)

Additional Living Books
- *Future Engineering: The Clean Water Challenge* by Robyn Friend
- *John Muir Wrestles a Waterfall* by Julie Danneberg
- *Extreme Earth: Waterfalls* by Patricia Corrigan

Notebooking (SCIDAT Logbook Information)

This week, you can have the students begin to fill out the Climate Sheet for the Swiss Deciduous Forest. They can also fill out the first part of their weather record sheets and the logbook sheets for groundwater and waterfalls. Here is the information they could include:

Climate Sheets
Area Map – Have the students color the region where Switzerland is found. Have the students put a star where Zurich is located. (*See map for answers.*)

Climate Information – This week, the students could include the following:
- ⇨ *Deciduous forests have warm summers, with temperatures averaging around 70°F, and cool winters, with temperatures dipping below freezing.*
- ⇨ *Deciduous forests typically have between thirty to sixty inches of rainfall per year.*

Interesting Facts – Answers will vary. Students can add the following facts:
- ⇨ *Deciduous forests are also known as temperate forests and are characterized by trees that shed their leaves during the fall season.*
- ⇨ *There are four distinct seasons in the deciduous forest – fall, winter, spring, summer. Each one lasts about three months each.*

Weather Record Sheets
Have the students record the weather from your area or from Switzerland over the week.

Earth Science Record Sheets
Groundwater
Information Learned
- ⇨ *When it rains, a fair amount of the water is absorbed into the soil. Just below the soil is a layer of porous rock that can also hold water and beneath that is a layer*

of impermeable rock that water cannot pass through. So the water sits in the porous rock, creating a layer of water known as an aquifer. The top of this layer is known as the water table.

⇨ *The water found in these aquifers is known as groundwater amd is a key source of fresh water. As this water soaks through the rock, it carve out huge caverns, underground rivers, and waterfalls by dissolving the minerals found in rocks like limestone.*

⇨ *When the porous rock layer meets the surface, such as on a hillside, a spring can form. A spring is when the groundwater bubbles out of the ground, forming a small stream or pool. This water is typically very pure and packed with minerals as it has filtered through the porous rock layer.*

Waterfalls
Information Learned

⇨ *A waterfall is a point in a stream or river when the water sudden falls downward.*

⇨ *This can be a short fall of a few feet or a much larger one. The tallest known waterfall is Angel Falls in Venezuela, in which the water falls 3,212 feet.*

⇨ *A waterfall typically occurs when a stream or river flows from hard rock to softer rock. The flowing water wears down the softer rock faster, which creates a drop where it falls.*

⇨ *Waterfalls can also be formed when the land is changed by a major event, like an earthquake or a landslide, which changes the land around the stream or river.*

⇨ *The most common way to classify a waterfall is by how the water falls. For example, a block waterfall is water that falls in a sheet from a wide river. Niagara Falls is an example of a block waterfall.*

Vocabulary

Have the older students look up the following term in the glossary in the Appendix on pp. 109-110 or in a science encyclopedia. Then, have them copy the definition onto a blank index card or into their SCIDAT logbook.

 **Aquifer** – An underground layer of porous rock that can hold water.

Scientific Demonstration: Groundwater Filter

Materials
- ☑ Plastic bottle
- ☑ Cotton balls
- ☑ Gravel
- ☑ Sand
- ☑ Soil
- ☑ Duct tape
- ☑ Water

Procedure

1. Cut the water bottle in half. Flip the top half of the bottle, place it inside the other half so that the top-opening is about an inch above the bottom, and use the tape to secure it in place.
2. Have the students stuff the cotton balls into the top opening and then cover that with gravel, sand, and then dirt.
3. Next, have the students mix several teaspoons of leftover soil and sand with about half a cup of water. Then, have them pour this dirty water into their filter and observe what happens.

Explanation

The purpose of this demonstration is to give the students a crude look at how the ground filters and cleans the water. The students should see dirty water seep down through the layers and come out of the bottom much cleaner than it began. The same process happens as the water filters through ground and fills the aquifers.

Take it Further
Have the students test different materials and sizes of rocks to see how that affects their results.

Multi-Week Projects and Activities

Multi-week Projects
✂ **Weather Poster** — Have the students add to their weather poster this week. (*Weather template pictures can be found in the Appendix on pg. 8.*)

Activities For This Week
✂ **Groundwater** — Have the students learn more about what groundwater is by watching the following video:
- https://www.youtube.com/watch?v=oNWAerr_xEE

✂ **Waterfall erosion** — Have the students see how a waterfall can erode the rock and create a pool. You will need a sponge, a bar of soap, like Ivory, and access to a sink. Turn the faucet on and place the sponge directly under the stream of water. Then, set the bar of soap on top of the sponge and let it sit there with the water falling on the same place for about ten minutes. Turn the water off and observe the hole that has been created in the soap bar.

✂ **Types of Waterfalls** — Have the students learn about the different types of waterfalls using the information from this website:
- http://www.scienceforkidsclub.com/waterfalls.html

Memorization

Copywork/Dictation
☞ **Copywork Sentence**
A waterfall is a point in a stream or river when the water sudden falls downward.

☞ **Dictation Selection**
When it rains, a fair amount of the water is absorbed into the soil. Just below the soil is a layer of porous rock that can also hold water. Beneath the porous rock is a layer of impermeable rock that water cannot pass through. The water sits in the porous rock, creating a layer of water known as an aquifer. The water found in these aquifers is known as groundwater, which is a key source of fresh water.

Notes

Chapter 17: Tracking Down Bogdanovich

Chapter Summary

The chapter opens with Agent Wuthritch making the hand-off with Yuroslav, when suddenly the monitors in the surveillance van go blank. Agent DeBlose quickly realizes that he is dealing with a mole and they head down to the river to follow Yuroslav and figure out who the traitor is. As they speed down in the Kayak X, Agent DeBlose tells them about the river they are traveling on. The reach the dam and spot the mole and Yuroslav head inside. They drive the Kayak X into the nearby lake, learn about the lake, and enter the dam through the other side in hopes of surprising the villain. The twins and their expert are captured and forced to go over the dam in the Kayak X, which thankfully has been equipped with a grappling hook. They find their way back to the dam, but in the end, the mole and Yuroslav get into a waiting helicopter, taking the vial of poison with them to Siberia. The chapter ends with the twins and Agent DeBlose being forced off the dam once more, but this time they are not in the safety of the Kayak X!

Supplies Needed

Demonstration	Projects and Activities
• Flour, Aluminum pan, Eye dropper, Water	• Materials will vary based on how the students choose to represent the three stages of a river.

Optional Schedule for Two-Days-A-Week

Day 1	Day 2
☐ Read the section entitled "River Reconnaissance" of Chapter 17 in *SSA Volume 4: Earth Science*.	☐ Read the section entitled "Lakeside Lasso" of Chapter 17 in *SSA Volume 4: Earth Science*.
☐ Fill out an Earth Science Record Sheet on SL pg. 85 on rivers.	☐ Fill out an Earth Science Record Sheet on SL pg. 86 on lakes.
☐ Add facts to the Climate Information on SL pg. 81.	☐ Add weather to the Weather Information Sheet on SL pg. 82; Go over the vocabulary word and enter it into the Earth Science Glossary on SL pg. 100.
☐ Add weather to the Weather Information Sheet on SL pg. 82.	☐ Do the demo entitled "River Erosion"; write information learned on SL pg. 88.
☐ Do the copywork or dictation assignment and add it to the Earth Science Notes on SL pg. 88.	☐ Work on one or all of the multi-week activities.

Optional Schedule for Five-Days-A-Week

Day 1	Day 2	Day 3	Day 4	Day 5
☐ Read the section entitled "River Reconnaissance" of Chapter 17 in *SSA Volume 4: Earth Science*.	☐ Read the section entitled "Lakeside Lasso" of Chapter 17 in *SSA Volume 4: Earth Science*.	☐ Read one or all of the assigned pages from the encyclopedia of your choice; write narration on the Earth Science Notes Sheet on SL pg. 88.	☐ Read one of the additional library books.	☐ Do the copywork or dictation assignment and add it to the Earth Science Notes sheet on SL pg. 88.
☐ Fill out an Earth Science Record Sheet on SL pg. 85 on rivers.	☐ Fill out an Earth Science Record Sheet on SL pg. 86 on lakes.	☐ Do the demo entitled "River Erosion"; write information learned on SL pg. 88.	☐ Go over the vocabulary word and enter it into the Earth Science Glossary on SL pp. 100.	☐ Add weather to the Weather Information Sheet on SL pg. 82.
☐ Add facts to the Climate Information on SL pg. 81.	☐ Add weather to the Weather Information Sheet on SL pg. 82.		☐ Choose one of the activities for the week to do; fill out the project record sheet on pg. 90.	☐ Work on one or all of the multi-week activities.

Science-Oriented Books

Living Book Spine
- Chapter 17 of *The Sassafras Science Adventures Volume 4: Earth Science*

Optional Encyclopedia Readings
- *Basher Science Planet Earth* pg. 56 (River), pp. 58-59 (Lake)
- *Discover Science Weather* (No pages scheduled.)
- *Usborne Children's Encyclopedia* pp. 24-25 (Following a River)
- *Usborne Encyclopedia of Planet Earth* pp. 122-123 (Rivers), pp. 124-125 (River Erosion)

Additional Living Books
- *One Well: The Story Of Water On Earth* by Rochelle Strauss
- *Planet Earth: Rivers and Lakes* by Rani Iyer
- *The World Around Us: Rivers* by Cecilia Minden
- *Rivers, Lakes, and Oceans (The Restless Earth)* by Gretel H. Schueller

Notebooking (SCIDAT Logbook Information)

This week, you can have the students begin to fill out the Climate Sheet for the Swiss Deciduous Forest. They can also fill out the first part of their weather record sheets and the logbook sheets for rivers and lakes. Here is the information they could include:

Climate Sheets
Interesting Facts – Answers will vary. Students can add the following fact:
⇨ A stream is a fast moving body of freshwater much smaller and shallower than a river. Some you can even walk across. Many streams flow together to form a river.

Weather Record Sheets
Have the students record the weather from your area or from Switzerland over the week.

Earth Science Record Sheets
Rivers
Information Learned
⇨ A river is a moving body of water. Rivers carry water from their source, or beginning, towards a larger body of water, such as a lake or ocean.
⇨ The source of a river is typically found high up in the mountains, where lots of smaller streams come together to eventually form the river.
⇨ Rivers have three key stages:
  1. The upper stage is where the river's course is steep and the water moves very fast. This is where many of the streams are joining in to create the river. The water in this stage is very clear.
  2. The middle stage is where the bed of the river smooths out and the path of the river begins to flatten. The river widens in this stage and makes wide loops, known as meanders. Smaller rivers may also join in at this point to create a larger river.
  3. The lower stage is where the river flattens out to join the lake or ocean. The water in this stage is pack with sediment, so it appears very muddy. At this point, the river may also split into channels,

to form what is known as a delta.
- *Rivers not only carry water, they also pick up loose sand, soil, mud, and rocks. They carry this material along, which causes the bed, or bottom, of the river to become wider and deeper.*

LAKES
INFORMATION LEARNED
- *A lake is a barely moving body of water that is not connected to the ocean. In other words, it is land-locked.*
- *Water flows into lakes from streams and rivers and can also flow out via streams or rivers.*
- *Lakes can form as the river finds its way into a basin and fills the area with water before continuing its journey downward.*
- *Lakes can also form when a bend in the river narrows and is eventually basically cut off by the deposit of mud and rock carried by the river.*
- *Lakes can also be man-made; these lakes are dug out by humans and known as reservoirs.*
- *Lakes do lose water due to evaporation, so they need to be continually fed by water or they will dry up.*
- *Lakes are typically filled with freshwater, but there are several saltwater lakes. Lakes become salty when there is no outlet for the water. Instead, the water evaporates, leaving behind the minerals it contained and making the remaining water salty.*

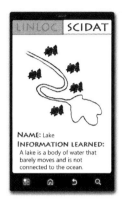

VOCABULARY
Have the older students look up the following term in the glossary in the Appendix on pp. 109-110 or in a science encyclopedia. Then, have them copy the definition onto a blank index card or into their SCIDAT logbook.
- EROSION – The slow and steady wearing down of soil or rock by wind or water.

SCIENTIFIC DEMONSTRATION: RIVER EROSION
MATERIALS
- ☑ Flour
- ☑ Aluminum pan
- ☑ Eye dropper
- ☑ Water

PROCEDURE
1. Pour about two cups of flour into the aluminum pan. Have the students shape the flour into a mountain near one end of the pan.
2. Have the students use the eye dropper to slowly add water to the top of their flour mountain. Keep adding water until they have created several "river courses" on the flour mountain and pools at the base (around a cup of water).

EXPLANATION
The purpose of this demonstration is to give your students a chance to see how a river forms from beginning to end. The students should see that the water moves faster at the top of their mountain and much slower at the base, just like a river in nature.

TAKE IT FURTHER
Have the students repeat the demonstration with different shapes of flour mounds to see the many different river courses they can create.

Multi-Week Projects and Activities

Multi-week Projects
✂ **Weather Poster** – Have the students add to their weather poster this week. (*Weather template pictures can be found in the Appendix on pg. 8.*)

Activities For This Week
✂ **River Diagram** – Have the students make a diorama, poster, or salt dough map showing the three stages of a river that they learned about.

✂ **Lake** – Take a field trip to a local lake or pond.

Memorization

Copywork/Dictation
☞ **Copywork Sentence**
A river is a moving body of water.

☞ **Dictation Selection**
A river, which is a moving body of water, has three key stages. The upper stage is where the rivers course is steep and the water moves very fast. The middle stage is where the bed of the river smooths out and the path of the river begins to flatten. The lower stage is where the river flattens out to join the lake or ocean.

Notes

Chapter 18: The End of Earth Science

Chapter Summary

The chapter opens with the twins and the local expert, Agent DeBlose, using their Swiss Army Knife parachutes to safely float down to the river's edge. The head back to the Triple S Headquarters, where the twins find a place alone and open the LINLOC app on their phones, finding out it is time to zip back to the Left-Handed Turtle Market. They land in a recycling bin, where their phones receive the bonus data on caring for Earth's resources. Uncle Cecil and President Lincoln then let the twins in on the second purpose of the recycling bin—a roving anti-dog tank. We then switch to the Man with No Eyebrows, where we learn that although he has erased Summer's memory, he has not found satisfaction. His next plan is to find and abduct some of the villains the twins have dealt with, erase their memories, and train them to be his army. Back at the recycling bin/anti-dog tank, the four spot Summer walking absentmindedly around on Pecan St. Mrs. Pascapali approaches her and they find out that Summer can't remember a thing. Uncle Cecil's neighbor invites her to stay, while the twins, Cecil, and President Lincoln head back to the lab. They quickly figure out what the Man with No Eyebrows had done to Summer and Cecil remembers that he has a canister with Summer's thoughts stored somewhere in his lab. The book ends with Blaine and Tracey wondering if the canister could be found and Summer's memory restored.

Supplies Needed

Demonstration	Projects and Activities
• Recycling bins	• No Supplies Needed

Optional Schedules for Two-Days-A-Week

Day 1	Day 2
☐ Read the section entitled "Bonus Data" of Chapter 18 in *SSA Volume 4: Earth Science*.	☐ Read the section entitled "Jumbled Geography" of Chapter 18 in *SSA Volume 4: Earth Science*.
☐ Add the bonus data information onto the Earth Science Notes Sheet on SL pg. 91.	☐ Review the vocabulary found on SL pp. 95-100.
☐ Do the demo entitled "Recycling Plan"; write information learned on SL pg. 92.	☐ Do the copywork or dictation assignment and add it to the Earth Science Notes sheet on SL pg. 92.
☐ Go over the vocabulary word and enter it into the Earth Science Glossary on SL pg. 100.	☐ Work on one or all of the multi-week activities.

Optional Schedule for Five-Days-A-Week

Day 1	Day 2	Day 3	Day 4	Day 5
☐ Read the section entitled "Bonus Data" of Chapter 18 in *SSA Volume 4: Earth Science*. ☐ Add the bonus data information onto the Earth Science Notes Sheet on SL pg. 91.	☐ Do the demo entitled "Recycling Plan"; write information learned on SL pg. 92. ☐ Go over the vocabulary word and enter it into the Earth Science Glossary on SL pg. 100.	☐ Read the section entitled "Jumbled Geography" of Chapter 18 in *SSA Volume 4: Earth Science*. ☐ Review the vocabulary found on SL pp. 95-100.	☐ Do the copywork or dictation assignment and add it to the Earth Science Notes sheet on SL pg. 92. ☐ Read one of the additional living books from your library; write narration on the Earth Science Notes Sheet on SL pg. 91.	☐ Work on one or all of the multi-week activities.

Science-Oriented Books

Living Book Spine
- 📖 Chapter 18 of *The Sassafras Science Adventures Volume 4: Earth Science*

Optional Encyclopedia Readings
- *Basher Science Planet Earth* pg. 122 Conservation
- *Discover Science Weather* (No pages scheduled.)
- *Usborne Children's Encyclopedia* pp. 46-47 (Useful Earth), pg. 52 (Helping our Planet)
- *Usborne Encyclopedia of Planet Earth* pp. 22-23 (Earth's Resources)

Additional Living Books
- 📖 *10 Things I Can Do to Help My World* by Melanie Walsh
- 📖 *Earth's Resources (Gareth Stevens Vital Science: Earth Science)* by Alfred J. Smuskiewicz
- 📖 *Where Does the Garbage Go?: Revised Edition (Let's-Read-and-Find Out About Science)* by Paul Showers and Randy Chewning
- 📖 *The Adventures of a Plastic Bottle: A Story About Recycling* by Alison Inches and Pete Whitehead
- 📖 *The Three R's: Reuse, Reduce, Recycle* by Nuria Roca and Rosa M. Curto

Notebooking (SCIDAT Logbook Information)

This week, you can have the students fill the Earth Science Notes sheets with the bonus data. Here's the information they could include:

Bonus Data
Resources and Recycling
- ⇨ As humans, our technological advances can be hurting our planet.
- ⇨ Our need for paper requires the cutting down of lots of trees.
- ⇨ Our need for food, and other goods, has changed the landscape through farming and mining.
- ⇨ Our need for electricity and mass produced products has created pollution and trash that is damaging our environment.
- ⇨ To help protect our natural resources, we can:
    1. Recycle things like paper and plastic.
    2. Throw our trash away in a trash can, not on the ground.
    3. Plant things, either in our own backyards or as part of a service project in our communities.
    4. Switch off lights and other electronics when we are not using them.
    5. Walk or ride our bikes when we are not traveling very far.

Vocabulary
Have the older students look up the following term in the glossary in the Appendix on pp. 109-110 or in a science encyclopedia. Then, have them copy the definition onto a blank index card or into their SCIDAT logbook.

- **RECYCLING** – The act of turning waste, like used cans, bottles, and paper, into new, usable materials.

Scientific Demonstration: Recycling Plan

Materials
- ☑ Recycling bins

Procedure
1. Work with the students to create a recycling plan for each family. You can visit the following website for more information on recycling and ideas on how to set up a recycling plan.
   - http://greenbootcamp.blogspot.com/2007/12/week-two-setting-up-at-homerecycling.html
   - http://www.bravenewleaf.com/environment/2007/11/how-to-set-up-a.html

Multi-Week Projects and Activities

Multi-week Projects
- ✂ REVIEW – Review what the students have learned this semester.

Activities For This Week
- ✂ EARTH'S RESOURCES – Have the students make a collage of the Earth's resources. Have them search for pictures in magazines of the different things that Earth produces that we use, such as food, trees, and oil. Then, have them cut out the pictures, label them, and glue them to a poster labeled "Earth's Resources." You can have your older students label the resources with where they can be found as well.

Memorization

Copywork/Dictation

☞ **Copywork Sentence**

Recycling is the act of turning waste, like used cans, bottles, and paper, into new, usable materials.

☞ **Dictation Selection**

Recycling is the act of turning waste, like used cans, bottles, and paper, into new, usable materials. Recycling helps to reduce the amount of trash left on Earth.

Notes

Appendix

LAB REPORT SHEET

Title
_____

Hypothesis (What I Think Will Happen)
_____
_____
_____

Materials (What We Used)

    _____   _____

    _____   _____

    _____   _____

    _____

Procedure (What We Did)
_____
_____
_____
_____
_____
_____
_____
_____
_____
_____

Observations and Results (What I Saw and Measured)
_____
_____
_____

Conclusion (What I Learned)

Microscope Worksheet

Sample Name
_____

Description of What I Saw

_____
_____
_____
_____
_____
_____
_____
_____
_____
_____
_____

Sketch of What I Saw

Magnification Power _____x   Magnification Power _____x

MICROSCOPE WORKSHEET

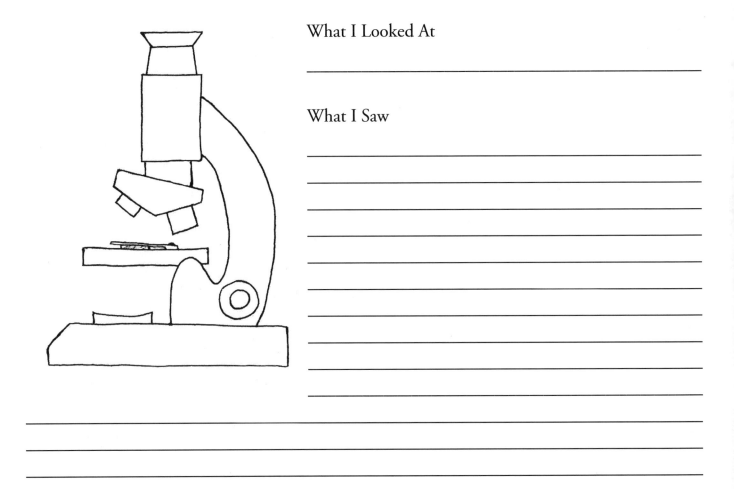

What I Looked At
_____

What I Saw
_____
_____
_____
_____
_____
_____
_____
_____
_____
_____
_____
_____

My Drawing

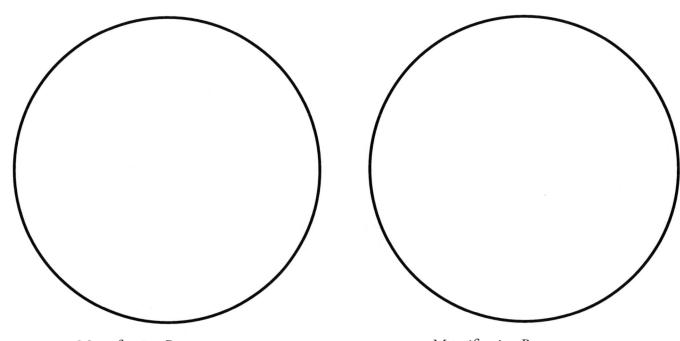

Magnification Power _____ x   Magnification Power _____ x

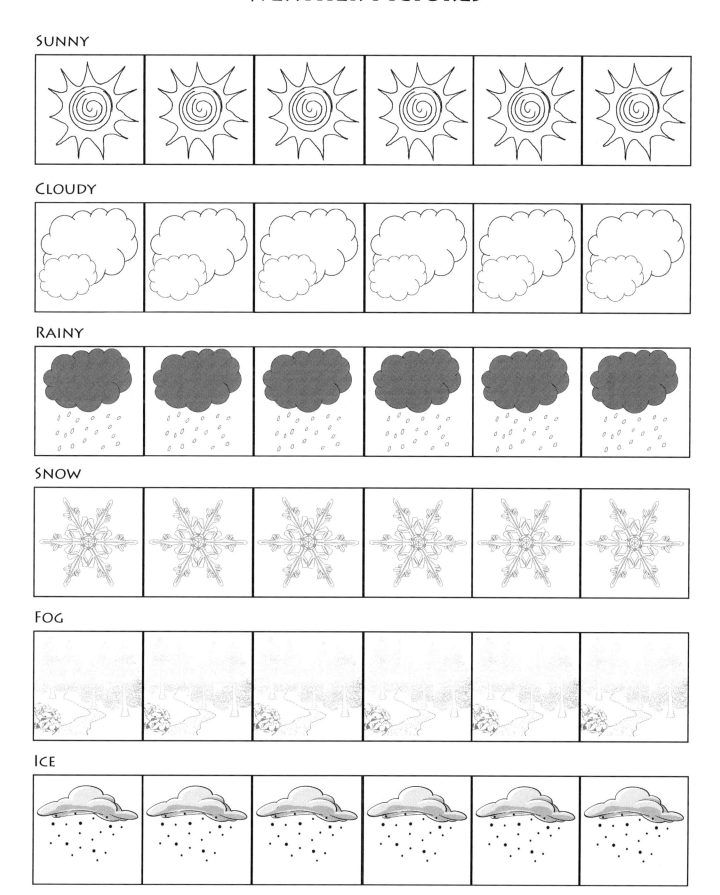

THE WATER CYCLE

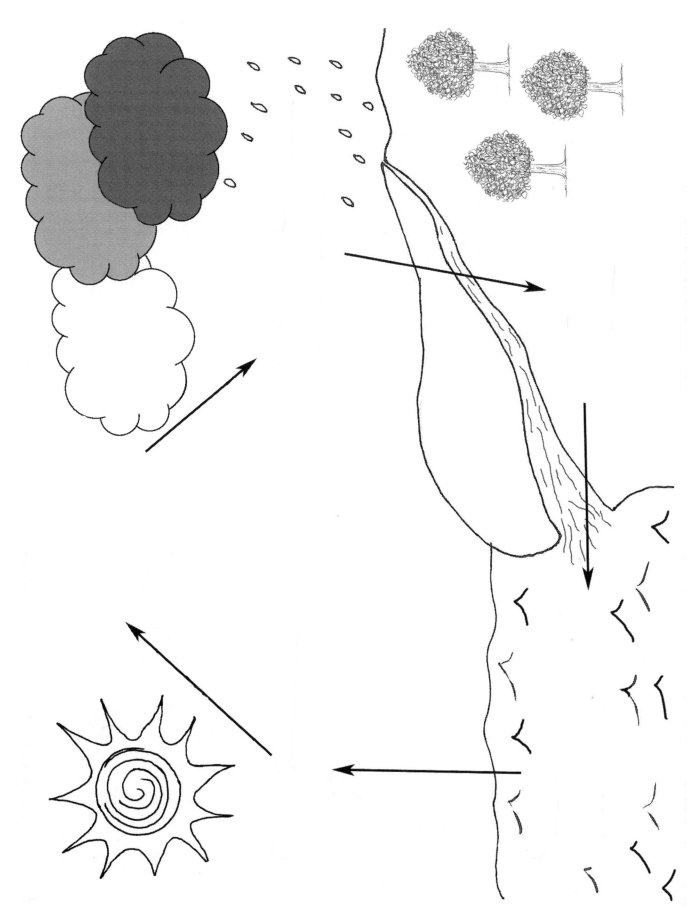

The Water Cycle

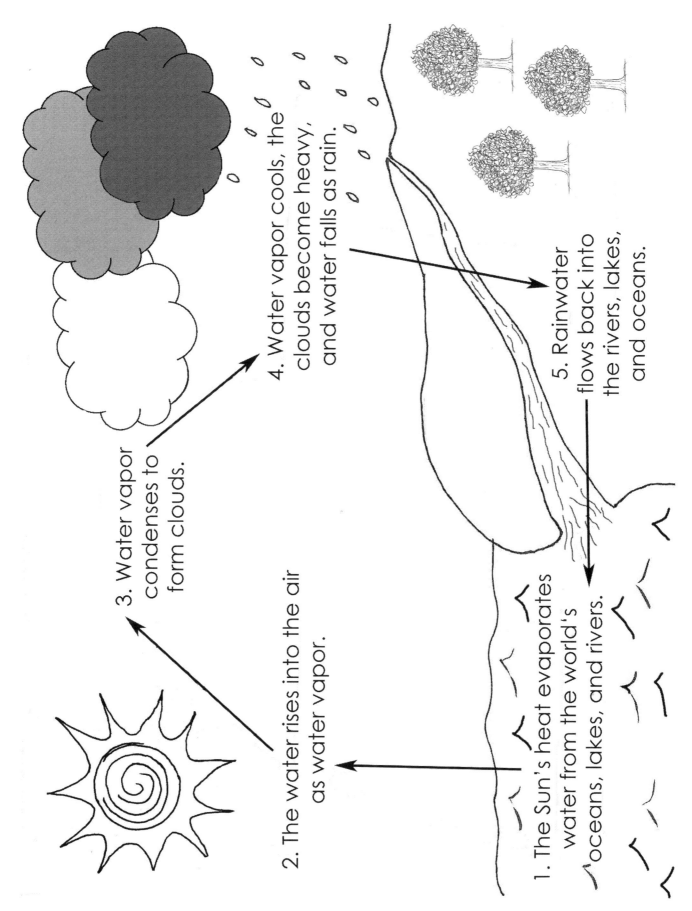

The Nitrogen Cycle

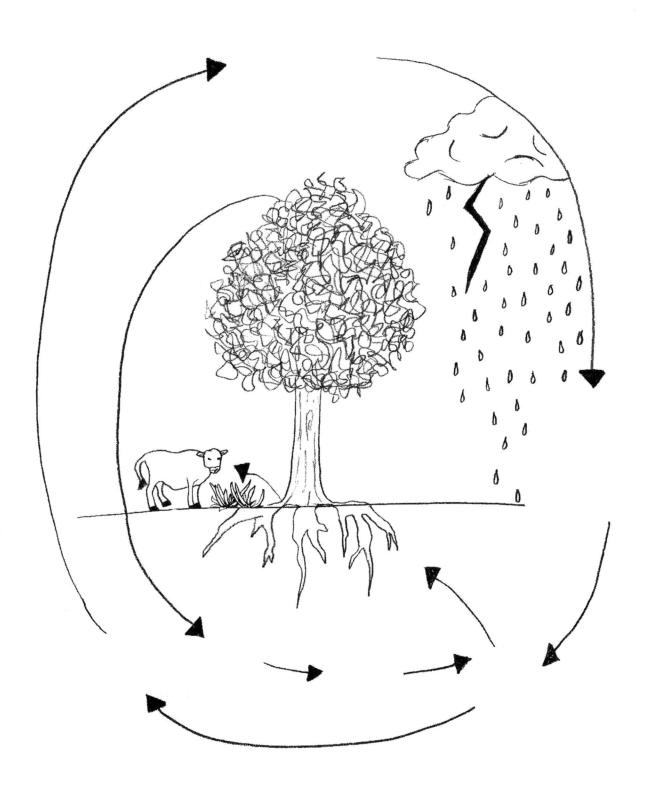

The Nitrogen Cycle

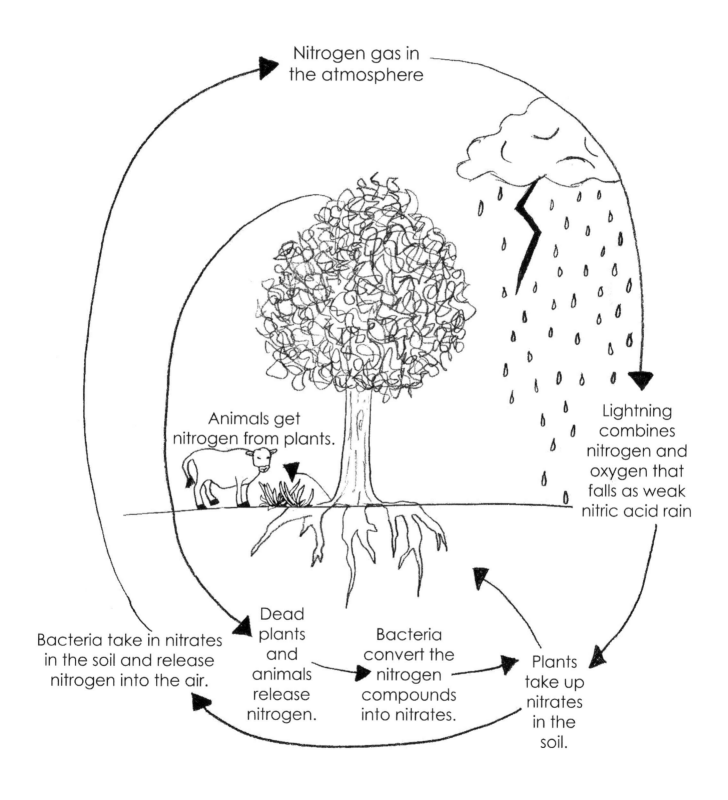

THE PHOSPHORUS CYCLE

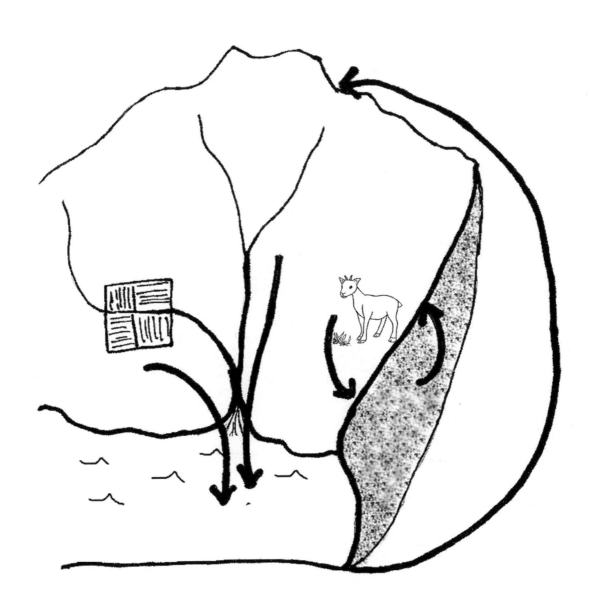

THE PHOSPHORUS CYCLE

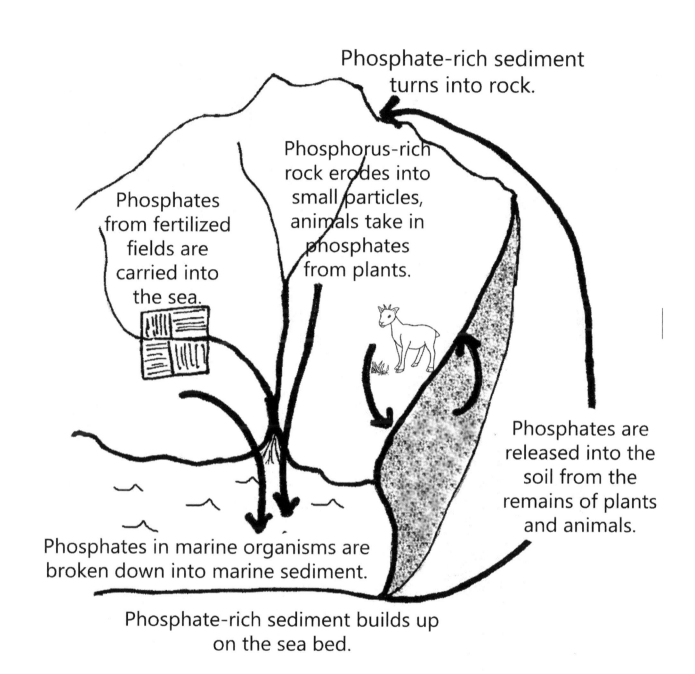

GLOSSARY

Earth Science Glossary

A

- AQUIFER – An underground layer of porous rock that can hold water.
- ATMOSPHERE – A layer of gas that surrounds the Earth.

B

C

- CLIMATE – The average weather in an area over a given period of time.
- CLOUDS – A collection of condensed water vapor.
- CORAL – A group of tiny animals with hard shells, similar to rock; they live together in a colony that builds upon itself.
- CURRENTS – The movement of water or air in a particular direction, usually due to a difference in temperature

D

- DROUGHT – A long period of without rain.

E

- EQUATOR – An imaginary line that divides the earth into the Northern and Southern Hemisphere.
- EROSION – The slow and steady wearing down of soil or rock by wind or water.

F

- FLOOD – An overflow of a large amount of water due to heavy rains or melting snow.
- FOG – A thick cloud of tiny water droplets that are suspended very close to the Earth's surface.

G

- GUST – A short burst of wind moving at a high speed.

H

I

J

K

L

M

- MONSOON – A season of strong winds and heavy rain.

N

- **NATURAL CYCLE** – A naturally occurring event that happens over and over again in the same order.

O

P

- **PRECIPITATION** – Rain, snow, sleet, or hail that falls to the ground.

Q

R

- **RECYCLING** – The act of turning waste, like used cans, bottles, and paper, into new, usable materials.

S

- **SANDSTORM** – A strong, violent winds that stir up loose sand and sediment, carrying it to another location.
- **SEASON** – A collection of days with a typical weather pattern.
- **SNOWFLAKE** – A collection of ice crystals that form a shape with six similar sides.

T

- **THUNDERSTORM** – A storm with thunder and lightning.
- **TORNADO** – A spinning funnel of wind that touches the ground and is also connected to the clouds above.

U

V

- **VAPOR** – Tiny droplets of water in the air.

W

- **WEATHER** – Conditions, like windy, cloudy, sunny, or rainy, that change daily.
- **WIND** – The movement of air in the atmosphere created by temperature differences.

X

Y

Z

QUIZZES

Earth Science Quiz Answers

Quiz #1
1. Warm, cold
2. B, C, A
3. D
4. B
5. F0, F5
6. C

Quiz #2
1. Tropical, temperate
2. Equator
3. Air, condenses, water vapor, micro-droplets, bigger, gravity
4. C
5. B
6. D

Quiz #3
1. Winter
2. A
3. C
4. below, refreezes
5. D
6. Fall, winter, spring, summer

Quiz #4
1. D
2. Center of the Earth (or equator), north, south
3. Answers can include: hot, dry, dusty, or no rain
4. C
5. B
6. A

Quiz #5
1. A
2. higher, less
3. C
4. D
5. B, D, C, A

Quiz #6
1. 5, 2, 4, 1, 3
2. D
3. True
4. No a true

Quiz #7
1. D
2. C
3. D
4.
5. Answers are on the map below.

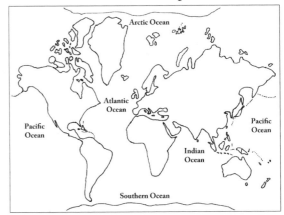

Quiz #8
1. Warm, cool
2. D
3. B
4. C
5. 3, 1, 2
6. A

Earth Science Quiz #1
Chapters 2 and 3

1. Temperate grasslands, like the prairies of Oklahoma have (warm / cold) summers and (warm / cold) winters.

2. Match the global wind pattern with its description.

   Trade winds _____

   Prevailing westerlies _____

   Polar easterlies _____

   A. These winds are found near the north and south poles. They blow up to the poles and curve from east to west.

   B. These winds are found near the equator. They flow north or south towards the equator and curve west due to the spin of the Earth.

   C. These winds are found in between the equator and the poles. They blow slightly towards the poles from the west to the east.

3. Wind is _____.

   A. The movement of atmospheric gases on a large scale

   B. Described by using two factors – speed and direction

   C. Caused by the uneven heating of the surface of the Earth

   D. All of the above

4. Tornadoes form _____.

   A. Normally in the winter months

   B. As two currents spiral and spin around each other

   C. From weak storms with light rain

   D. All of the above

5. The Fujitsu scale is used to describe the strength of a tornado. It ranges from F0 to F5, with _____ being the weakest tornado and _____ being the strongest tornado.

6. A downburst is _____ often associated with a thunderstorm.

   A. A weak downward current of air

   B. A strong upward current of air

   C. A strong downward current of air

   D. A weak upward current of air

Earth Science Quiz #2
Chapters 4 and 5

1. The two types of rainforests are _____ and

   _____.

2. Tropical rainforests, like the one found in the Congo, occur near the _____.

3. Fill in the blanks with the following words: air, water vapor, gravity, bigger, micro-droplets, and condenses.

   Rain forms when warm, moist _____ rises and _____

   to form a cloud of _____, the _____

   collect together to form _____ droplets which fall to the ground

   because of _____.

The Sassafras Guide to Earth Science ~ Quizzes

4. Monsoons are _____.

    A. The seasonal changes in the strongest winds of a region

    B. The cause of the wet and dry seasons that you find throughout the tropics

    C. Both A and B

    D. None of the above

5. A flash flood can occur after a period of _____.

    A. Little rain

    B. Intense rain

    C. Dry weather

    D. Heavy winds

6. A thunderstorm can include _____.

    A. Thunder and lightning

    B. Heavy rain

    C. High winds

    D. All of the above

Earth Science Quiz #3
Chapters 6 and 7

1. When it is summer in the northern hemisphere, it is _____ in the southern hemisphere, where Patagonia is found.

2. Wind chill is when _____.

   A. The temperature your body feels is colder than the actual temperature

   B. The temperature your body feels is warmer than the actual temperature

   C. The temperature your body feels is the same as the actual temperature

   D. None of the above

3. Snowflakes come in many shapes and sizes, but each one is ____-sided.

   A. 2

   B. 4

   C. 6

   D. 8

The Sassafras Guide to Earth Science ~ Quizzes

4. Freezing rain occurs when the air close the surface is (below / above) freezing,

so that the precipitation (melts / refreezes) when it hits the surface.

5. Frost quakes occur _____.

    A. In the middle of the night when the temperatures are at their coldest

    B. Because of a sudden rapid freezing of the ground

    C. Because of a shift of the earth's plates

    D. Both A and B

6. Name the four seasons.

    F_____

    W_____

    S_____

    S_____

Earth Science Quiz #4
Chapters 8 and 9

1. The Mongolian Desert _____.

    A. Is located far inland

    B. Is also known as the "land of the blue sky"

    C. Has extreme temperature changes

    D. All of the above

2. Hot deserts are found around the _____. Cold deserts are found _____ and _____ of the center of the Earth.

3. During a drought, the conditions are typically _____ _____.

4. An oasis is a _____.

    A. Place where a pool of water can be found

    B. Natural or man-made

    C. Both A and B

    D. None of the above

5. Sandstorms _____.

    A. Contain lots of rain and wind

    B. Have strong violent, winds that stir up sand and loose sediment

    C. Arrive with lots of warning

6. The day/night cycle is caused by _____.

    A. The rotation of the Earth

    B. A curtain of clouds that come in and out

    C. None of the above

Earth Science Quiz #5
Chapters 10 and 11

1. At high altitudes, the temperature is typically much cooler because _____.

    A. There is less air to trap the heat

    B. There is more air to trap the heat

    C. The wind is less

    D. None of the above

2. As you get (lower / higher) in altitude, the effect of gravity is (less / more).

3. The atmosphere is _____.

    A. A blanket of air that surrounds and protects the planet

    B. Composed of the troposphere, stratosphere, mesosphere, thermosphere, and exosphere

    C. Both A and B

    D. None of the above

4. Clouds _____.

   A. Are composed of dust and water vapor or ice

   B. Form when warm air holding water vapor cools down

   C. Can be flat, wispy, or puffy

   D. All of the above

5. Match the type of cloud with its description.

   Cirrus clouds _____   A. Large puffy, clouds that grow vertically between the layers

   Alto clouds _____   B. Thin, wispy clouds found high in the atmosphere

   Stratus clouds _____   C. Low, flat clouds that often block out the sun

   Cumulus clouds _____   D. Mid-level clouds that can be both puffy and flat

Earth Science Quiz #6
Chapters 12 and 13

1. Put the steps of the water cycle in order and label them on the cycle picture below.

    _____ Rainwater and snow melt flow back into the rivers, lakes, and oceans.

    _____ The water rises into the air as water vapor.

    _____ As the water vapor cools, the clouds become heavy and water falls as rain or snow.

    _____ The sun's heat evaporates water from the world's oceans, lakes, and rivers.

    _____ Water vapor condenses to form clouds.

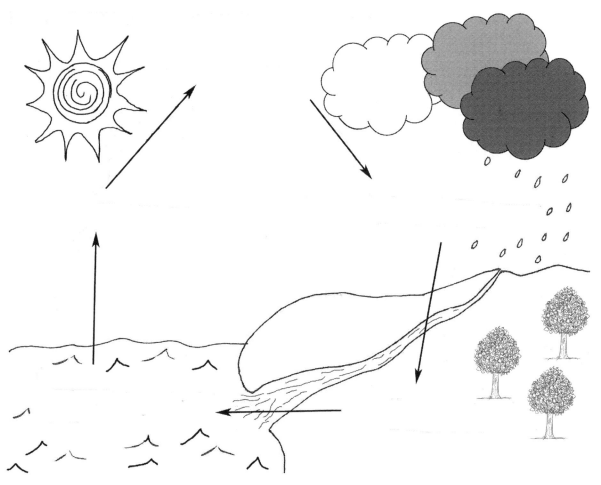

The Sassafras Guide to Earth Science ~ Quizzes

2. Fog _____.

   A. Is a thick cloud of tiny water droplets that are suspended very close to the Earth's surface.

   B. Forms in the same way that clouds do.

   C. Typically only happens when the air is very humid or full of water vapor.

   D. All of the above

3. The nitrogen cycle is a (true / not a true) cycle explaining how nitrogen moves between plants, animals, bacteria, the air, and the soil.

4. The phosphorus cycle is a (true / not a true) cycle explaining how phosphorus moves between plants, animals, water, and the soil.

Earth Science Quiz #7
Chapters 14 and 15

1. On Earth, you can find _____ water.

    A. Fresh

    B. Salt

    C. Brackish

    D. All of the above

2. Coral reefs are home to the _____ mix of plants and animals found in the oceans.

    A. Least

    B. Average

    C. Greatest

3. Currents in the ocean are produced by _____.

    A. Wind on the surface

    B. Differences in temperature

    C. The spinning motion of the Earth

    D. All of the above

4. Hurricanes are also known as _____.

   A. Typhoons

   B. Cyclones

   C. Both A and B

   D. None of the above

5. Label the oceans on the map below with the following - Arctic Ocean, Atlantic Ocean, Indian Ocean, Pacific Ocean, and Southern Ocean.

Earth Science Quiz #8
Chapters 16 and 17

1. Deciduous forest have (warm / cool) summers and (warm / cool) winters.

2. A deciduous forest _____.

    A. Is also known as a temperate forest.

    B. Contains trees that shed their leaves during the fall season

    C. Has four distinct seasons

    D. All of the above

3. A waterfall is a point in a stream or river when the water sudden falls _____.

    A. Upward

    B. Downward

    C. Sideways

    D. None of the above

4. Groundwater collects in _____.

    A. Aquifers

    B. Porous rock

    C. Both A and B

5. Match the stage of the river with its description.

    Upper stage _____   1. The bed of the river smooths out and the path of the river begins to flatten.

    Middle stage _____   2. The river flattens out to join the lake or ocean.

    Lower stage _____   3. The rivers course is steep and the water moves very fast.

6. A lake is a barely moving body of water that _____ connected to the ocean.

    A. Is not

    B. Is

Made in the USA
San Bernardino, CA
05 March 2020